Peek-A-Blue berry farm

A collection of Classic blueberry recipes from across the United States
compiled by Chelsey Keeler

ISBN 1-4276-0081-3
Copyright © 2006 Peek-A-Blue Berry Farm
All Rights Reserved

First Printing - July 2006
Second Printing - August 2006
Third Printing - October 2006
Fourth Printing - November 2006
Fifth Printing - April 2007
Sixth Printing - July 2007

Published by Aardvark Global Publishing Company, LLC

9587 So. Grandview Dr.   1-800-614-3578
Salt Lake City, UT 84092 USA   www.AardvarkGlobalPublishing.com

INTRODUCTION

In 2002 my twelve-year-old daughter took it upon herself to send a letter to "Country Magazine". Her request, she wanted a blueberry recipe from every state to add to her collection. We had started a blueberry farm in the country hills of Steuben County, New York State in 1999. Chelsey was notified around August that she was selected to have her request published in the December/ January 2003 edition. She was also warned that you might get a lot of mail, so be prepared. The magazine was released for circulation and the letters started rolling in, ten or fifteen a day, and then twenty five to fifty a day, and around seventy five to eighty per day. Readers from every one of the United States and three countries opened up their recipe boxes, hearts, and favorite ways of using blueberries. Overwhelmed with mail and cookbooks, gifts and cards, special people made this become so much fun. Our family, friends and neighbors would call to see "how many letters did Chelsey get today?!". Even the mailman, started leaving special mail bundles, inquiring how many was she up to. After an estimated 2,500 recipes and over 1,000 letters, we soon learned that people are very generous and they love blueberries. We also learned that there is a real desire for people to bond to a common something given the chance, and in this case, it was blueberries and a twelve-year-old girl. With this wonderful response, "Country Extra Magazine" in the May 2004 edition featured "how she was helped" as a follow up to the huge response from people all across America, and even three foreign countries. Explaining her desire to create a cookbook with her blueberry recipes received, more letters continued to follow and even now in May 2006, a letter or two still comes her way. Letters and recipes signed "Grama" or "your friend", "Chef", "school teacher", "retired", "Blueberry Associations", all shared their fondest memory of "blueberry days" or "blueberry recipes" (usually intertwined together). She has been blessed to be a part of many lives that have the common bond of blueberries.

Prior to starting our three-acre blueberry farm, for over 15 years we had a few plants in our side yard. Around mid July, blueberries became the coveted food group. So much so, our dog, "Chance" would eat them right off the bush and her two cats, Thomasina and Chloe assisted in the "supervision" of the picking, making them also a part of the picking ritual.

Starting off, we hand dug over 1,000 holes and hand planted every plant. Our plants were ordered out of Arkansas after finding them on the Internet. They arrived Easter Sunday in 1999, and we had to give the venture a name. We agreed on Peek-A-Blue Berry Farm and it seemed appropriate. Our farm overlooks a beautiful five-acre pond, and is nestled in among her Grandparents family dairy farm. Now with over 2,200 plants, family, friends and neighbors enjoy the large blueberries. We also supply a local grocery store chain, Wegmans of Corning, New York with the in season "home grown" Peek-A-Blue berries.

This cookbook that my now sixteen-year-old daughter typed on her own during the summer of her fourteenth year, is ready to share with our family, friends and blueberry lovers from all over. Thank you to all who have sent a favorite recipe, notes of encouragement, gifts, or just well wishes. We received so many duplicate recipes and due to allotment of space, every state is represented with many unique recipes, but not all recipes were able to be included. For all of the recipes listed in the book, thank you for sharing with us from your collection. For all of the recipes that did not get included, please understand how special and how much we appreciated your generosity, you are definitely not forgotten, perhaps in a second volume someday!

Acknowledgements

Now to my daughter, Chelsey (aka, Fred), I am so proud of you. On your own, you have brought so many people who love blueberries together in one collection. This first book that you have authored, is printed as my gift to you. As you use this cookbook for all the special blueberry times in your life, remember, you will always be the best gift life has given to me.

Love,
Da

Classic Blueberry recipes from across the United States

The Story of the Letters
By: Chelsey Keeler
Author of this cookbook, compiler of the recipes, and recipient of the letters

Actually it all started when my Grandma replied to a gentleman in "Country Magazine" telling him about the Oregon School House, a local one-room schoolhouse built in 1869. He later wrote back stating that he had gotten about 100 letters from interested readers on his requested topic. That seemed to be the ultimate for me; I didn't think I had gotten a total of 100 letters addressed to me in my whole life (all 12 years). Grandma then came up with the suggestion that I write in to "Country" inquiring about blueberry recipes from each state. I did, but then tried not to get my hopes up, there was no way 100 people would take time out of their hectic lives to write to a 12 year old.
I was never more wrong, as previously stated, the letters and packages simply poured in. One day I even caught myself being disappointed at only 20 letters in the mailbox. Then when the rush slowed down to about two a day, I stopped being overwhelmed and finally began to appreciate every single letter, all 1000 plus of them. Time and time again, I have marveled at how the two words " blueberry recipes" could bring back such powerful memories for people, and that they were so generous to include me in on them. It was wonderful to be drawn into someone's life and reminisce about how delicious Aunt Lena's coffee cake tasted or how serene it was to sit in the sun soaked fields with friends, picking the hours away.
To answer some of the questions people asked- we have High Bush blueberries plants; the types being Blue Crop, Duke, Collins, Northland, Blue Ray and Blue Jay.
Also as a little "disclaimer", much as I would like to have tried each recipe individually, all were not tested. They are written as they were received with no claim, so as I have done with so many of them, enjoy and have fun with as many as you can.
I have included excerpts of letters from each State and Country so that you may enjoy a few of the personal bits of information that were shared with me. I wish I could have printed them all. It was so interesting to hear everyone's personal stories involving blueberries. After receiving so many notes and recipes, it became evident that a collection of them should be shared with others from the ones that so generously shared with me.
 Included in one of the letters, it stated how Robert Frost once wrote a poem entitled " Blueberries". After reading it, I knew it had to be a part of the book, as it is so wonderfully expressed. "The blue's but a mist from the breath of the wind". Thanks so much for this opportunity, and I hope this cookbook is enjoyed by all.

Chelsey Keeler
Peek-A-Blue Berry Farm

8

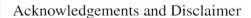

Acknowledgements and Disclaimer

The recipes contained in this collection and publication were selected by State and uniqueness of the recipe with its association to blueberries. In no manner does the author, publisher or any other entity, including Peek-A-Blueberry Farm claim ownership or exclusives rights to the submissions received. All credits for each recipe go directly to the originator(s), creator(s), or association(s) that may have discovered, created, or shared the recipe(s) for its original requested use; to be placed and published in a cook book collection by Chelsey Keeler; Peek-A-Blue berry farms. "Country" and "Country Extra" magazine are part of Reiman Publications and by their choice, printed the request of the author for receipt of blueberry recipes from every State in the USA, via USPS letter. Reiman Publications was not requested to endorse this publication, nor in any fashion is this publication to be considered the works, ownership or reflective of Reiman Publications.

Each recipe or recipes that are printed were selected by State as the primary criteria, uniqueness to blueberries, and considered a good representation to the collection of "Classic blueberry recipes from across the United States". The recipes were recorded and presented in "as received" letter, email electronic, or any other format for the contents, ingredients and intended use of the recipe. The author, Peek-A-Blueberry Farms, or the publisher of this book make no claims to the validity of the recipes, only that they were transferred to an electronic format for this publication. Any representation that is in reference to any other owned work, is not intended for infringement and in no way intended to take from, copy and/or distribute without permission, or publish without regards to the originator. In the event of an error in; judgment, transfer, ingredients, instructions, bake times, bake temperatures or print, corrections will be made to the next release of the book and at no time will the author, Peek-A-Blue berry farms or the publisher be required to replace or refund the selling cost of the book, or be liable for any other consequences from the use of the recipes or information contained in this publication..

As the intent of this collection is to share with others blueberry recipes, memories of incidents concerning blueberry activities, cultural differences and likeness concerning blueberries of all fifty United States, and three countries. It is with gratitude that all contributors; printed or not selected, are acknowledged for their generosity to make this collection a reality. For sharing with us, and allowing us to share with others, we thank you and hope you enjoy "Classic Blueberry recipes from across the United States".

*T*able of Contents

PEEK-A-BLUEBERRY FARM

Blueberry Sauce

Submitted by: Peek-A-Blueberry Farm
From: Bath, New York
Bake Time: Medium heat

Ingredients:
1/4 Cup of sugar   1/4 Cup of water   1 Tbls. of cornstarch
2 Cups of blueberries (fresh or frozen)

Instructions:
Place ingredients into sauce pan and simmer until mixture is thick. Serve on pancakes, ice cream and waffles.

Blueberry Pancakes

Submitted by: Peek-A-Blueberry Farm
From: Bath, New York
Bake Time: 2-5 minutes
Bake Temperature: 325 degrees

Ingredients:
1 Cup of pancake mix   1 cup of blueberries (fresh or frozen)
3/4 Cup of water

Instructions:
Mix water and pancake mix. Spoon onto griddle, about 1/4 cup per pancake. Sprinkle blueberries on top of each pancake. Cook until bubbles appear. Flip and cook on the other side, 1 to 2 minutes. Serve with syrup of blueberry sauce.

Blueberry Banana Loaf

Submitted by: Grandpa Bud Keeler
From: Peek-A-Blueberry Farm
Bake Time: 60 minutes
Bake Temperature: 350 degrees

Ingredients:

1 Cup whole wheat flour	1/2 Tsp. Cinnamon	3 Tbls. butter
3/4 Cup flour	1/4 Tsp. Salt	1 egg
1 Tsp. baking soda	1/3 Cup sugar	1 Cup mashed banana
1 Cup blueberries	1 Tbls. lemon juice	
1/2 Cup quick cooking rolled oats		

Instructions:

Preheat the oven to 350 degrees. Spray a 8 1/2 by 4 1/2 inch loaf pan with nonstick cooking spray. Combine both flours, baking soda, cinnamon and salt in a large bowl. Stir in the oats. Cream the butter and sugar in the large bowl of an electric mixer. Beat in the egg. Add the bananas and lemon juice. Stir until well blended. Add the dry ingredients and mix just until moistened. Gently fold in blueberries. Pour the batter into the loaf pan and bake for about an hour, or until inserted toothpick comes out clean. Let the bread cool in the pan for 10 minutes. Turn out on a wire rack to cool completely. Wrap and refrigerate several hours before cutting into 16 slices.

Lemon Blueberry Squares

Submitted by: Grandma Sylvia Loghry
From: Peek-A-Blueberry Farm
Bake Time: 50 minutes
Bake Temperature: 350 degrees

Ingredients:

3/4 Cup butter	1/4 Tsp. Salt	1/3 Cup flour
1/2 Cup confectioners' sugar	2 Cups sugar	1 Cup blueberries
2 Tsp. vanilla	2 1/4 Cups flour	2 Tsp. grated lemon rind
6 eggs	1/4 Cup confectioners' sugar	1/2 Cup lemon juice

Instructions:

Heat oven to 350 degrees. Line a 13 by 9 by 2 inch baking pan with aluminum foil. Coat with nonstick cooking spray. In a bowl, stir together butter, 1/2-cup confectioners' sugar, vanilla and salt. Gradually stir in 2 1/4 cups flour until smooth. Press dough over bottom of prepared pan. Bake at 350 for 20 minutes. In a large bowl, mix 2 cups sugar and 1/3-cup flour. Whisk in eggs until smooth. Stir in lemon rind and juice. Fold in berries. Pour filling over crust. Bake at 350 for 30 minutes. Let cool in pan on wire rack. Dust with remaining confectioners' sugar. Cut into squares and serve.

Best Blueberry Stuff

Submitted by: Chelsey Keeler & Grandma Sylvia Loghry
From: Peek-A-Blueberry Farm
Bake Time: 35 minutes
Bake Temperature: 375 degrees

Ingredients:
Crumbs:

1 ½ Cups Quick cook Oatmeal	1 ½ Cups All purpose flour	1 cup brown sugar
1 Cup butter – softened	1 tsp. baking soda	

Filling:

4 cups blueberries	1 ½ cups sugar	3 tbsp. Cornstarch
1 tsp. baking soda	¼ cup water	3 0z. Grape gelatin

Instructions:
Put filling ingredients in saucepan and cook together until thickened. Add gelatin, stir into mixture and cool. Mix crumb ingredients in a bowl and spread half of crumb mixture evenly in a 9 X 13 pan. Pour filling mixture on crumbs, then sprinkle the rest of the crumbs on the filling. Bake at 375 degrees for 35 minutes.

Berry Blondies

Submitted by: Grandma Sylvia Loghry
From: Peek-A-Blueberry Farm
Bake Time: 50 minutes
Bake Temperature: 325 degrees

Ingredients:

2 eggs	1 Tsp. baking powder	1 Cup raspberries
2/3 Cup sugar	1 Tsp. Vanilla	1 Cup blueberries
1 1/3 Cups flour	1/2 Tsp. salt	
6 Oz. white chocolate, chopped	5 Tbls. unsalted butter, cut into pieces	

Instructions:
Heat oven to 325. Line a 9 by 9 by 2 inch pan with foil, grease. In metal bowl set over saucepan, barely simmering, melt white chocolate and butter, stirring until smooth. Let cool to room temperature. In a large bowl, beat eggs, sugar, vanilla on medium speed until thickened and pale, 3 minutes. On low, gradually beat in chocolate mixture. Into a bowl, sift flour, baking powder, and salt. Beat into chocolate mixture until combined. Spread evenly into prepared pan. Sprinkle with berries. Bake at 425 until top is very lightly browned but center is still soft when pressed, 45 to 50 minutes.

Lemon Blueberry Tart

Submitted by: Grandma Sylvia Loghry
From: Peek-A-Blueberry Farm
Bake Time: 20 minutes
Bake Temperature: 350 degrees

Ingredients:

18 cream sandwich cookies   1 1/2 Cups blueberries   5 Tbls. yellow cornmeal
1/4 Cup bottled lemon curd   1 egg   1/2 Cup whipped topping
1 Tbls. grape jelly

Instructions:

Heat oven to 350. Coat 11 by 7 by 1 inch rectangular tart pan with removable bottom with nonstick cooking spray. Break apart cookies. Place in food processor; add cornmeal. Process until cookies are finely crushed. Add egg; process until crumbs stick together. Between 2 sheets of plastic wrap, roll dough into 12 by 8 inch rectangle. Press into tart pan. Bake at 350 or until lightly browned around edge, 15 to 20 minutes. Transfer pan to wire rack, let cool. With a sharp knife, release the edge of crust from sides of pan. Transfer to serving platter. In a small bowl, stir together whipped topping and lemon curd. In a small saucepan, heat grape jelly until melted. Remove from heat. Add blueberries to jelly. With metal spatula, spread lemon filling over crust. Top with berries. Serve immediately or refrigerate loosely covered for up to 2 hours.

Blueberry Shake

Submitted by: Peek-A-Blueberry Farm
From: Bath, New York

Ingredients:

2 Cups blueberries   2 Tbls. lemon juice   2 small bananas
1 1/2 Cups vanilla yogurt   1/3 Cup honey   1 Cup vanilla ice cream

Instructions:

Combine blueberries, bananas, honey and lemon juice and blend on high in blender. Add yogurt and ice cream; blend until thick and smooth. Serve immediately in cold glasses decorated with sprigs of mint.

Blueberry Drop Cookies

Submitted by: Grandma Sylvia Loghry
From: Peek-A-Blueberry Farm
Bake Time: 10-12 minutes or until golden brown
Bake Temperature: 375 degrees

Ingredients:

1 cup shortening
1 Tsp. vanilla extract
1 Tsp. salt
Additional sugar

2 cups sugar
2 Tsp. baking powder
1/4 Cup Milk
4 Cups plus 1 Tbsp. all purpose flour (divided)

4 eggs
1 Tsp. baking soda
2 Cups fresh blueberries

Instructions:

In a large mixing bowl, cream shortening and sugar. Add eggs, one at a time, beating well after each addition. Add vanilla; mix well. Combine 4 cups flour, baking powder, baking soda and salt; add to the creamed mixture alternately with the milk. Coat the blueberries with the remaining flour; gently fold into the batter.Drop by tablespoonfuls 2 in. apart onto greased baking sheets. Sprinkle with additional sugar. Bake at 375 for 10-12 minutes or until golden brown. Remove to wire racks to cool. Yield: 7 dozen.

Blueberry Summer Pie

Submitted by: Grandma Sylvia Loghry
From: Peek-A-Blueberry Farm
Bake Time: 50 minutes
Bake Temperature: 375 degrees

Ingredients:

4 Cups peaches
2/3 Cup sugar
3 Tbls. flour
1 Tsp. vanilla
1 large egg beaten with 1 Tbls. water

1 Cup blueberries
1 Tbls. grated orange rind
1 Tsp. Cinnamon
1 cup flour

3 Tbls. cold water
1/4 Tsp. salt
3 Tbls. olive oil

Instructions:

Heat oven to 375. Coat a 10-inch pie pan with non-stick cooking spray. In a large bowl, gently stir together the peaches, blueberries, sugar, orange rind, cinnamon, vanilla, and 3 tablespoons flour until fruit is evenly coated. In a separate bowl, stir together the cup of flour, salt and olive oil. Add the cold water, 1 tablespoon at a time, tossing with a fork until the mixture together in a ball. Form the dough into a disk. Place between two sheets of waxed paper; with a rolling pin, roll into 11- inch round, turning dough occasionally to get an even thickness. Remove top of waxed paper. Crimp the edges of the rust all around. Spoon the filling into the prepared pie dish. Invert the crust into your hand; peel off the waxed paper. Invert crust on top of pie filling so crimped edge is facing up. Brush the crust evenly with the egg mixture. Cut a few vents in the top crust. Bake at 375 for 45 to 50 minutes. Transfer pie to a cooling rack Cool completely.

Peek-A-Blue berry farm

ALABAMA

Blueberry production, mostly of the rabbiteye type, is increasing in Alabama. There are over 34 species of blueberries worldwide. The rabbiteye (Vaccinium ashei) is native to the Southeast and is the most popular species grown throughout Alabama. The blueberry belongs to the same family (but a different genus) as the wild huckleberry. Huckleberries have 10 very hard seeds, and their berries are blackish when ripe. Blueberry fruits are larger and have many small, softer seeds that are not very noticeable when eaten. Most blueberries produce fruits which have a powdery gray-blue "bloom" on the surface of the skin which helps reduce moisture loss after harvest.

Alabama Cooperative Extension

Blueberry Crunch

Submitted by:	Jan Simmons
From:	Duncanville, AL.
Bake Time:	30 minutes
Bake Temperature:	350 degrees

Ingredients:

1 large can crushed pineapple
2 Cups blueberries
1 Cup chopped pecans

1 box yellow cake mix
1 1/2 sticks butter

1/4 Cup sugar
3/4 Cup sugar

Instructions:

Lightly grease a 9 by 13 inch pan and spread undrained pineapple. Add blueberries. Sprinkle 3/4-cup sugar over berries. Spread dry cake mix on berries, and then drizzle melted butter all over cake mix. Add pecans and sprinkle 1/4-cup sugar on top. Bake 30 minutes at 350.

Blueberry Lemonade

Submitted by: Gloria Cole
From: Hueytown, AL

Ingredients:

2 Cups blueberries   1/2 to 3/4 Cup of fine sugar (pending berry sweetness)
1 Cup fresh squeezed lemon juice   2 Cups of cold water

Instructions:

Taste one of the blueberries to determine sweetness. In a blender or food processor, process the blueberries and superfine sugar until smooth. If blueberries are frozen, thaw first. Strain the puree through a fine mesh strainer into a medium sized bowl, stirring and pushing on the puree to get all the juice. Pour the blueberry juice into a 1quart container. Add the lemon juice and water and stir or shake to combine. Check for sweetness, adding superfine sugar to taste, if needed. Chill until very cold and serve over ice. Makes 4 servings.

Blueberry Dessert

Submitted by: Mary Kemplin
From: Montgomery, AL.
Bake Time: 20 minutes
Bake Temperature: 350 degrees

Ingredients:

11 graham crackers crushed   1 Cup sugar   3 eggs, well beaten
8 Oz. softened cream cheese   1/4 Cup melted butter   20 Oz. canned blueberries

Instructions:

Mix cream cheese, 1/2 cup sugar and eggs in a bowl. Prepare graham cracker crust by mixing melted butter, 1/2 cup sugar and crushed graham crackers. Pat out in 9x13 cake pan. Pour cheese mixture over crust. Bake 15-20 min., remove from oven, spoon the blueberries sauce over top. Chill until set.

Blueberry Crumble

Submitted by: Glen M. Ellis
From: Blountsville, AL.
Bake Time: 30 minutes
Bake Temperature: 350 degrees

Ingredients:

(2) 16 Oz. packs of frozen blueberries
1 Tbls. Lemon juice   1/3 Cup sugar   1/2 Cup all purpose flour
1/2 Cup brown sugar, packed   1/2 Cup butter
1/2 Cup quick cooking oats, uncooked

Instructions:

Combine blueberries, sugar and lemon juice, place in buttered 8" baking pan, combine oats, flour, and brown sugar. Cut in the butter until mixture resembles course oatmeal. Spread over fruit mixture and bake at 350 degrees for 30 minutes. Serve warm plain or over ice cream.

Blueberry Congealed Salad

Submitted by: Betty Faulkner
From: Auburn, AL

Ingredients:

2 small boxes grape Jell-O   2 Cups hot water   1 Tsp. vanilla
1 can blueberry pie filling   1/2 Pint sour cream   1/2 Cup sugar
8 Oz. softened cream cheese   1 Cup chopped nuts
1 large can crush pineapple, undrained.

Instructions:

Mix Jell-O and hot water. Add pie filling and pineapple. Let congeal overnight. Combine cream cheese, sour cream, sugar and vanilla in a separate bowl. Spread over top of congealed mixture. Sprinkle with nuts. Cut into squares and serve.

Blueberry Pound Cake

Submitted by: Millie Taylor
From: Monroeville, AL
Bake Time: 60 minutes
Bake Temperature: 350 degrees

Ingredients:

1 box butter cake mix	3 eggs	1/2 Cup Crisco oil
1 15 Oz. can blueberries drained	8 Oz. package of cream cheese	1 Cup chopped nuts

Instructions:

Mix cake mix, oil, cream cheese and eggs with mixer. Fold in Blueberries and nuts. Spoon into well greased and floured tube or bundt pan. Bake at 350 degrees for 45 to 60 minutes. Let set in pan for 10 minutes. Remove and finish cooling.

Blueberry Pie

Submitted by: Ruth Branum
From: Owens Cross Roads, AL.

Ingredients:

2 graham cracker pie shells	1 box-powdered sugar	1 can blueberry pie filling
4 8 Oz. package cream cheese	1 Cup crushed pecans	
1 small container of Cool Whip		

Instructions:

Cream together cream cheese and powdered sugar until all lumps are out. Stir in pecans. Spread in 2 Pie shells, top with 1/2 cool whip for each making a "bowl" in the middle of each pie shell. Top each with 1/2 can of pie filling and refrigerate.

Blu-Bana Bread

Submitted by: Annie Dabney
From: Fitzpatrick, AL
Bake Time: 50 minutes
Bake Temperature: 350 degrees

Ingredients:

1 Cup butter	2 Tsp. vanilla	3 Tsp. allspice
2 Cups sugar	5 medium bananas, mashed	2 Tsp. soda
4 eggs	4 Cups flour	1 Tsp. baking powder
1/2 Tsp. salt	2 Cups of blueberries	

Instructions:

Preheat oven to 350 degrees. Grease and flour two 5x9 inch loaf pans. Cream together butter and sugar. Beat in eggs and add vanilla. Fold in bananas and 2 cups of flour. Reserve 2 Tbls. of flour to coat blueberries. Sift together the remaining flour, allspice, soda, baking powder and salt, fold into banana mixture. Sprinkle the 2 Tbls. of flour on to the blueberries, coat well and fold into batter. Divide the batter into loaf pans. Bake for approximately 50 minutes. Test with toothpick in center to determine when done. Yields 2 loaves.

Alma'a Blueberry Pie

Submitted by: Joan Payne
From: Altoona, AL

Ingredients:

1 can Eagle Brand milk	1/2 Cup pecans (optional)	1/4 Cup lemon juice
1 small can of Cool Whip	1 can blueberries, drained	1 graham cracker pie crust

Instructions:

Mix Eagle Brand milk and lemon juice until thick. Add cool whip and nuts. Work blueberries in last. Put into piecrust. Refrigerate and chill, serve cold.

Blueberry Salad

Submitted by: Ida Graham
From: Anniston, AL.

Ingredients:

1/2 Cup sugar
1/2 Cup chopped nuts
1/2 Tsp. vanilla
(2) 3 oz. packages blackberry gelatin

(1) 8 oz. package cream cheese
1 can blueberries, drained
(1) 18 oz. can crushed pineapple, drained

2 Cups boiling water
1/2-pint sour cream

Instructions:

Dissolve gelatin in boiling water. Drain pineapple and blueberries saving the juice. Add enough water to make 1 cup liquid. Add gelatin mixture and stir in fruit. Pour into a 2-quart dish. Cover and refrigerate until firm. Combine cream cheese, sugar, sour cream and vanilla; spread over salad. Sprinkle nuts on top. Refrigerate. Serves 12.

Blueberry Pie

Submitted by: Mildred Stephenson
From: Hartselle, AL.
Bake Time: 45 minutes
Bake Temperature: 450 degrees

Ingredients:

1 Quart blueberries
butter
(1) 9" pie shell, top and bottom

1 Tbls. lemon juice
5 Tbls. flour

1 Cup sugar

Instructions:

Wash blueberries, toss with sugar and flour mixture. Place in 9" pie shell. If any sugar mix is left, sprinkle over berries. Dot with butter, add lemon juice. Place crust on top, vent and bake at 450 degrees until brown. Serve warm with ice cream.

Blueberry Salad

Submitted by: Wyatt Bayer
From: Mobile, AL.

Ingredients:

2 small boxes of grape Jell-O   1 large can of blueberry pie filling
2 Cups boiling water   1 Cup chopped pecans
1 large can of crushed pineapple   1/2 Cup sugar
1 small tub sour cream   8 Oz. cream cheese

Instructions:

Mix together Jell-O, water, pineapple, blueberry pie filling and 1/2 cup chopped pecans. Allow mixture to congeal. Mix together cream cheese, sour cream and sugar. Spread mixture over top of congeal salad. Top with remaining crushed pecans. Chill and serve.

Sour Cream & Berry Pie

Submitted by: Mrs. Leslie Ableman (Susan Brinskele, Petaluma, Calif.)
From: Clanton, AL.
Bake Time: 8 minutes
Bake Temperature: 375 degrees

Ingredients:

1 Cup graham cracker crumbs   1/2 Cup sugar   1 Tbls. vanilla
1/4 Cup chopped pecans or walnuts 3 Tbls. cornstarch   3 Cups blueberries
2 Tbls. Flour   1 Tsp. unflavored gelatin 1 Tbls. sugar
1 1/3 Cups milk   1/3 Cup butter, melted   1 1/2 Cups sour cream

Instructions:

For crust: combine cracker crumbs, pecans or walnuts, flour, 1 Tbsp. sugar. Stir in melted butter. Toss to mix well. Press mixture onto bottom and up sides of a 9 inch pie plate. Bake at 375 oven for 8 minutes. Cool on wire rack. For filling: in a medium sauce pan combine 1/2 cup sugar, cornstarch, and gelatin, stir in milk. Cook and stir till thickened and bubbly; cook and stir for 2 minutes more. Place sour cream in medium bowl. Gradually stir milk mixture into the sour cream, stir in vanilla. Cover and chill for 1 hour, stirring once or twice. Stir berries into sour cream mixture. Turn filling into cooled crust. Cover and chill, garnish for at least 6 hours or up 24 hours. Garnish with fresh fruit.

Blueberry upside-down pudding

Submitted by: Higgins Legal
From: Birmingham, AL.
Bake Time: 45 minutes
Bake Temperature: 375 degrees

Ingredients:

1 Pint blueberries	1 egg	1/2 Tsp. salt
3/4 Cup sugar	1 Cup flour	1/2 Cup milk
1/4 Cup shortening	1 1/2 Tsp. baking powder	

Instructions:

Pour washed blueberries into buttered casserole (1-1/2 qt. Size) and sprinkle with 1/4 cup sugar. Cream shortening, 1/2 cup sugar and 1 egg together. Sift together flour, baking powder, and salt. Add this to creamed mixture alternately with 1/2 cup milk. Pour batter over blueberries. Bake at 375 degrees for 45 minutes. Spoon out while hot and serve with cream, whipped cream, ice cream, or milk.

Serving Blueberries

Submitted by: Alabama Cooperative Extension System
From: State of Alabama

Instructions:

1. Top cereal with blueberries, or sprinkle fresh blueberries on ice cream, melons, or meringue shells.
2. Use blueberry sauce on waffles, vanilla pudding or ice cream.
3. Stir blueberries into pancake, waffle or cake batter.
4. Make blueberries crepes. Place sweetened blueberries sprinkled with lemon juice in the center of freshly cooked crepes and roll the crepes. Top with whipped cream, more berries and blueberry sauce.

ALASKA

" We have a lot of blueberries in Alaska. Blueberry picking is something to look forward to every summer… We live in the bush in South East Alaska. We are 25 miles from the nearest town by boat. There are wild blueberry bushes all around our home. Many mornings we go out to pick berries and then make this blueberry pancake recipe."

" There are lots of recipes for baked blueberry desserts, but entrée recipes are not as common so I thought this might be a bit different for you. It makes a good dinner with crusty bread, corn on the cob and a tossed green salad. Don't substitute flounder or sole for the halibut. Their flesh is more delicate and they don't do well. Good luck with your project."

Blueberry Oatmeal Bread

Submitted by: Kathy Simpson
From: Anchor point, AK.
Bake Time: 60 minutes
Bake Temperature: 350 degrees

Ingredients:

2 Cups flour   1 Tsp. salt   1 Cup buttermilk
1 Tsp. baking powder   1/2 Tsp. Nutmeg   1/2 Cup light brown sugar
1 Tsp. baking soda   1 Cup oatmeal   1-1/2 Cups blueberries
1 Cup chopped nuts   1/3 Cup shortening   2 eggs

Instructions:

Sift all dry ingredients together. Cream shortening and sugar in separate bowl, add eggs & buttermilk. Gradually add dry ingredients and mix well. Add blueberries and nuts. Bake for 1 hour at 350 degrees.

Halibut With Blueberry Sauce

Submitted by: Neil Koeniger
From: Anchorage, AK
Bake Time: 40 minutes
Bake Temperature: 350 degrees

Ingredients:

1/2 Cup blueberries, rinsed   1 Tsp. sugar   1 Tsp. lemon juice
1/2 Cup sherry   2 Pounds halibut fillets, rinsed and patted dry

Instructions:

Stir first four ingredients together in a small saucepan. Bring to boil. Reduce heat and simmer 10 minutes. Place halibut fillets, skin side down, in an 8 by 11 by 2 baking pan. Bake, uncovered in 350-degree oven for 15 minutes. Pour sauce over fish. Bake 15 minutes longer. Serves 4-6.

Alaska Blueberry Shortcake

Submitted by: Debbie Poore
From: Homer, Alaska
Bake Time: 10 minutes
Bake Temperature: 450 degrees

Ingredients:

1 recipe baking powder biscuits   3 Tbls. evaporated milk   1 egg
blueberries with a little sugar   Dollop of sour cream   3 Tbls. sugar

Instructions:

Add egg, sugar and milk to biscuit recipe. Mix and form into 6 or 8 flattened shortcakes, or roll out dough and cut with larger cutter, can or glass. Bake on cookie sheet 8-10 minutes. Mix berries with a little sugar. Top shortcake with berries and a dollop of sour cream.

Blueberry Cobbler

Submitted by: Terry & Diane Whittlesey
From: Anchorage, AK
Bake Time: 35 minutes
Bake Temperature: 375 degrees

Ingredients:

2 Cups of Blueberries   1 Tsp. vanilla   1/2 Cup milk
1 Cup of flour   1 Tsp. baking powder   2 Tbls. melted butter
1 Cup of sugar   1 egg

Instructions:

Line bottom of 6 by 9 inch or 9 inch round pan with blueberries. Mix all other ingredients evenly in separate bowl and pour over berries. Bake at 375 degrees for 35 minutes or until golden brown.

Blueberry Batter Cake

Submitted by: Lorna Morse
From: Anchorage, AK
Bake Time: 60 minutes
Bake Temperature: 350 degrees

Ingredients:

2 Cups blueberries   1 Tsp. baking powder   1 Cup boiling water
Juice of 1/2 of a lemon 1 Cup flour   1 Cup sugar
3/4 Cup sugar   1 Tbls. cornstarch   1/4 Tsp. salt
1/2 Cup milk   1/4 Tsp. salt   3 Tbls. butter

Instructions:

Line a well greased 8 by 8 by 2-inch pan with berries, sprinkle with lemon juice. Cream the butter & sugar; add milk alternately with flour, baking powder and 1/4 teaspoon salt that has been sifted together. Spread over the berries. Combine sugar, cornstarch and remaining salt. Sprinkle over the cake. Pour boiling water over all. Bake at 350 degrees for 60 minutes. Serve with whipped cream or ice cream.

Alaska Blueberry Dessert

Submitted by: Marje Jensen
From: Pedro Bay, AK

Ingredients:

1 1/2 Cups sugar   1 1/4 Cups cold water   3 Tbls. lemon juice
1/2 Cup cornstarch   1 Cup blueberries   5 Cups blueberries
1/2 Tsp. salt   4 Tbls. butter   11 Oz. cream cheese
2 Cups graham cracker crumbs 3 Cups Cool Whip   1 stick butter, melted
2 Cups confectioners' sugar

Instructions:

Blend sugar, cornstarch, and salt in cold water and 1 cup of blueberries. Bring to a boil and cook until thick. Stir in 4 Tbsp. Butter and lemon juice, add 5 cups of blueberries, mix and let cool. Meanwhile, mix together graham cracker crumbs and 1 stick melted buttering a   9 by 13 inch pan, press evenly and put in freezer to firm up. Mix together confectioners' sugar and cream cheese, add Cool Whip. Layer over top of crust, spoon berries over top and let set for 1 hour refrigerated. Cut into squares.

Sour Cream Blueberry Pancakes

Submitted by: Ione Lynn
From: Petersburg, AK

Ingredients:

1 Cup flour   1 Tbls. sugar   1/4 Cup sour cream
3 Tsp. baking powder   1 large egg   1/4 Tsp. salt
1 Cup 2% low fat milk   1/2 to 1 Cup blueberries
2 Tbls. butter or margarine (melted)

Instructions:

Sift together flour, baking powder, salt, and sugar. Beat together egg, milk, and sour cream. Pour milk mixture over dry ingredients and blend with rotary beater until batter is just smooth. Stir in butter. Fold in blueberries. Pour 2 tablespoons batter onto hot griddle for each cake. Brown one side until golden. Turn and brown the other side. If cakes brown too fast, lower heat.

Bluebarb Jam

Submitted by: Carla Carlisle
From: Soldotna, AK

Ingredients:

3 Cups finely cut rhubarb   7 Cups sugar   1 bottle Certo (or 2 packs)
3 Cups crushed blueberries (about 5 cups uncrushed)

Instructions:

Mix fruit and sugar and put over high heat. Bring to a boil and then boil 1 minute. Stir constantly. Remove from heat and add Certo. Skim and put jam in hot jars. Put hot lids on jars and turn jars upside down for 15 minutes to seal.

Blueberry Buckle

Submitted by: Emily Smola and Abby Smola
From: Soldotna, AK
Bake Time: 35 minutes
Bake Temperature: 350 degrees

Ingredients:

1 Cup sour milk   1 1/2 Cups blueberries   1/4 Cup sugar
1 Tsp. vanilla   1 Tsp. soda   2 Tbls. butter
1 egg   2 1/2 Cups flour   1 1/2 Cups brown sugar
1 Tsp. salt   2/3 Cup oil   1/4 Cup brown sugar
1/2 Tsp. cinnamon   1/2 cups sliced almonds or pecans

Instructions:

Mix oil and 1 1/2 cups brown sugar. Add egg, sour milk and vanilla. Blend in flour, salt, soda. Fold in blueberries and pour into greased 9x13 pan. Combine white sugar, 1/4 cup brown sugar, cinnamon, butter and nuts. Sprinkle on top of batter. Bake at 350 degrees for 35 minutes.

Alaska Blueberry Pie

Submitted by: Marje Jensen
From: Pedro Bay, AK

Ingredients:

1 Tsp. grated lemon peel	2/3 Cup water	1 Tbls. lemon juice
1 Cup sugar	1 Tbls. Butter	1 small carton Cool Whip
2 1/2 Tbls. Cornstarch	2 1/2 Cups blueberries	1 pre-baked pie shell

Instructions:

Mix lemon peel, sugar, cornstarch, water, 1/2 cup blueberries, butter and lemon juice in a sauce pan and cook until very thick, stirring frequently. Remove from heat. Fold in 2 cups remaining berries. Pour into baked pie shell that has Cool Whip spread up the sides and across the bottom. Refrigerate 1 hour or until firm and set up. Serve chilled and garnish with fresh blueberries.

ARIZONA

"I'm 90 ½ years old, 10 days older than Arizona, but still like to help someone when I can. I started working for M.P. Railroad in Omaha Nebraska in 1928 at $1.85 per day and retired here in Phoenix in 1974 as M.P. State Rep."

"I'm sending you a recipe from Arizona which is our family's favorite. When we lived in Allentown, Pennsylvania, we had 4 blueberry bushes behind our home and had blueberries from June to September. We now live in Arizona and do not have access to very fresh blueberries..."

Cranberry Blueberry Pie

Submitted by: Anonymous
From: Tucson, AZ
Bake Time: 50 minutes
Bake Temperature: 400 degrees

Ingredients:

2 Cups cranberries   1/3 Cup flour   milk
2 Cups blueberries   1/2 small orange   1 1/4 to 1 1/2 Cups sugar
2 Tbls. unsalted butter, cut into bits   prepared pie shell for double-crusted 9" pie

Instructions:

Preheat oven to 400. Coarsely grind orange in a blender or processor. Combine this with the cranberries, then mix with blueberries. Toss this mixture well with sugar and flour. Spoon into a prepared pie shell. Dot butter bits on top. Adjust to pastry to fit, pressing down to release air. Seal and crimp sides. Brush top crust with milk and sugar. Cut steam vents on top and place on a cookie sheet. Bake for 45-50 minutes until crust is golden.

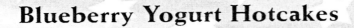

Blueberry Yogurt Hotcakes

Submitted by: Patricia Bristol
From: Tucson, AZ

Ingredients:

1 Tsp. salt
1/2 Tsp. baking soda
1/4 Tsp. ground cinnamon

1 large egg, lightly beaten
8 Oz. plain low fat yogurt
1 Cup fresh blueberries

1 1/4 Cups flour
1 Cup milk

Instructions:

Combine the dry ingredients and stir with a spoon. Stir in beaten egg, yogurt and milk until the batter is moistened. Fold in blueberries. Spray a non-stick griddle with cooking spray; heat over moderate flame. When griddle is hot, spoon batter onto griddle at 1/4 cup at a time. Turn hotcakes when top is bubbled. Makes 18 three-inch hotcakes, 55 calories each.

Blueberry Noodle Pudding

Submitted by: Patricia Bristol
From: Tuscan, AZ
Bake Time: 45 minutes
Bake Temperature: 350 degrees

Ingredients:

1/2 Pound wide noodles
4 Tbls. granulated fructose
Salt

8 Oz. vanilla yogurt
1 Cup pot cheese
1 Cup fresh blueberries

2 eggs
Cinnamon

Instructions:

Cook and drain noodles. Stir together eggs, pot cheese, yogurt, sugar and salt. Layer noodles, berries and cheese mixture in a 9-inch square nonstick cake pan. Sprinkle with cinnamon. Bake, uncovered, in a preheated 350 degree oven, 45 minutes until custard is set and top is lightly browned. Makes 10.

Bronco Bread

Submitted by: Sue Bridge
From: Cottonwood, AZ
Bake Time: 70 minutes
Bake Temperature: 325 degrees

Ingredients:

2 Tbls. soft butter
1/4 Cup hot water
1/2 Cup orange juice
1 1/2 Tbls. grated orange rind

1 egg beaten
1 Cup sugar
1/4 Tsp. soda
2 Cups flour

1/2 Tsp. salt
1 Cup blueberries
1 Tsp. baking powder

Instructions:

Combine butter, water, orange juice, orange rind, and mix well. Add egg and beat well. Add all other dry ingredients and mix. Fold blueberries in gently. Pour into greased loaf pan and bake at 325 for 1 hour and 10 minutes. Let cool before you cut it.

Dutch Blueberry and Cream Pie

Submitted by: Dale Darling
From: Peoria, AZ
Bake Time: 45 minutes
Bake Temperature: 400 degrees

Ingredients:

1 unbaked 9 inch pie shell
4 Cups blueberries
2/3 Cup sugar

1 Cup heavy cream
1/4 Tsp. salt
1/4 Cup flour

1 Tsp. vanilla
1/2 Tsp. nutmeg

Instructions:

Fill pie shell with blueberries. Combine remaining ingredients and blend well. Pour over berries. Bake in preheated oven for 45 minutes or until top is lightly browned. Cool and then chill. Serve garnished with uncooked berries and rosettes of sweetened whipped cream.

Meg's Scones

Submitted by: Meg
From: Arivaca, AZ
Bake Time: 20 minutes
Bake Temperature: 400 degrees

Ingredients:

2 Cups whole wheat pastry flour   1/4 Cup butter   3 Tsp. baking powder
1/2 Cup dried or fresh blueberries 1/4 Cup sugar   1/3 Cup rich milk
1/2 Tsp. salt   1/3 Cup orange juice

Instructions:

Mix dry ingredients. Cut in butter with a pastry knife. Add blueberries. Make a well in the center of the mix and pour in milk and juice. Stir together to make a dough. Knead 10 times. Pat onto clean floured surface, making a circle about 1/2 inch thick. Cut into 8 "pie" wedges. Sprinkle with cinnamon sugar and place on a greased cookie sheet. Bake in preheated oven for 20 minutes.

Blueberry Graham Cracker Dessert

Submitted by: Linda Mc Guire
From: Concho, AZ

Ingredients:

2 Cups crushed graham cracker   1 Cup small marshmallows   4 Tbls. sugar
blueberry pie filling   2 Tbls. melted butter   nuts are optional
2 Cups whipping cream

Instructions:

Mix two cups crushed graham crackers, two tablespoons sugar and two tablespoons melted butter and put into a 13 by 9 dish. Then mix 2 cups of the whipping cream, two tablespoons sugar and 1 cup small marshmallows. Spread half of it over the first layer. Next spread a can of blueberry pie filling and then spread the remaining whip cream mixture. Top with graham cracker crumbs and nuts.

Blueberry Pie in a Cake Pan

Submitted by: Ray Kopke
From: Phoenix, AZ
Bake Time: 60 minutes
Bake Temperature: 450 degrees

Ingredients:

3 3/4 Cups flour	5 Tbls. Butter	1 1/2 Cups brown sugar
1 Tbls. Sugar	1 1/2 Cups rolled oats	3/4 Cup butter
1 Cup oil	4 1/2 Tbls. Milk	4 Tbls. tapioca
1 1/3 Cup sugar	2 Tbls. lemon juice	6 Cups blueberries
Cinnamon		

Instructions:

Crust: mix 3 cups flour, one tablespoon sugar, oil, and milk together until well blended. Pat into bottom and sides of 13 by 9 pan.

Filling: Toss together blueberries, 1 1/3 cup sugar, tapioca, lemon juice and 4-5 tablespoons butter. Spoon into crust and dot with butter.

Topping: Mix together rolled oats, 3/4 cup flour, 1 1/2 cup brown sugar and 3/4 cup butter. Sprinkle with cinnamon to taste. Spoon on top of blueberry mixture. Bake at 450 for 10 minutes and then reduce heat and bake at 350 for 40-50 minutes until bubbling.

Blueberry Tea Cake

Submitted by: Darlene Mindler
From: Lk Havasu City, AZ
Bake Time: 50 minutes
Bake Temperature: 375 degrees

Ingredients:

2 Cups flour	1/4 Cup butter	3/4 Cup sugar
2 Tsp baking powder	2 Cups blueberries	1 egg
1/2 Tsp. salt	3/4 Cup milk	

Instructions:

Beat sugar and butter. Add an egg and beat well. Add dry ingredients and milk. Mix well. Gently fold in blueberries and pour into a 9-inch square metal pan. Mix up a dry crumb mixture- (1/2 cup sugar, 1/2 cup flour, 1/2 teaspoon cinnamon, 1/4 cup butter) and sprinkle on top of batter. Bake at 375 for 40-50 minutes.

Blueberry Sorbet

Submitted by: Shelly Hightower
From: Avondale, AZ

Ingredients:
4 Cups blueberries   1 Cup frozen apple juice concentrate

Instructions:
In a food processor combine blueberries and apple juice concentrate and blend until liquefied. Pour into a 11 by 17 inch baking pan. Cover and freeze until firm around edges, about 2 hours. With a heavy spoon break frozen mixture into pieces. In a food processor place mixture and blend until smooth but not completely melted. Spoon into a 9 by 5 inch loaf pan, cover and freeze until firm. Serve with in a few days. Serves 6.

Wild Blueberry Crisp

Submitted by: Shelly Hightower
From: Avondale, AZ
Bake Time: 45 minutes
Bake Temperature: 325 degrees

Ingredients:
5 Cups blueberries   1 Cup diced peeled apples   1/2 Cup flour
1/4 Cup sugar   1/2 Cup light brown sugar   1/2 Cup pecans
1/2 Tsp. grated lemon rind   2 Tsp. cinnamon   1 Tsp. nutmeg
1/2 Cup rolled oats   3 Tbls. butter   1/8 Teaspoon salt

Instructions:
Filling: In a small bowl combine blueberries, sugar, lemon rind, and apples. Mix well and place in a well buttered 8 by 8 by 2 inch pan.
Crisp: In a medium sized bowl combine brown sugar, cinnamon, nutmeg, flour, pecans, oats and salt. Rub in the butter with your fingers until it resembles coarse crumbs. Spread evenly over filling. Bake for 45 minutes at 325. Serves 6.

Blueberry Buckle

Submitted by: Faye Duvall
From: Phoenix, AZ
Bake Time: 40 minutes
Bake Temperature: 350 degrees

Ingredients:

3/4 Cup sugar	2 1/3 Cups flour	1/2 Cup milk
1/2 Cup brown sugar	1/2 Cup butter	1/2 Tsp. Salt
1 Cup blueberries	1/2 Tsp. nutmeg	1 egg
2 Tsp. baking powder	1/2 Tsp. cinnamon	

Instructions:

Beat 3/4 cup sugar with 1/4 cup butter until fluffy. Then add 1 egg and mix well. Next add 2 cups of flour, 1/2 teaspoon salt, 2 teaspoons baking powder and 1/2 cup milk. Beat well. Fold in 1 cup fresh or un-thawed frozen blueberries. Spread mixture evenly in a greased 9 inch square pan and sprinkle with the streusel topping (1/4 cup softened butter, 1/2 cup brown sugar, 1/3 cup flour, 1/2 teaspoon cinnamon, 1/4 teaspoon nutmeg). Bake at 350 for about 40 minutes or until golden brown.

Blueberry Squares

Submitted by: Dawn
From: Sedona, AZ
Bake Time: 45 minutes
Bake Temperature: 350 degrees

Ingredients:

3 eggs	1 Tsp. vanilla	1/2 Cup butter
1 Cup shortening	4 Cups flour	1 Tsp. cinnamon
3 Cups sugar	4 Cups blueberries	

Instructions:

Beat eggs in large bowl; add shortening and 2 cups sugar. Mix thoroughly. Add vanilla and 1 cup flour. Beat well. Add remaining 2 cups of flour. Beat well. Batter will be thick. Spread in a greased 15 by 12 inch jellyroll pan. Spread berries on top. Make a topping by mixing the 1 cup of flour, 1 cup of sugar and the teaspoon of cinnamon together. Cut in the 1/2 cup of butter and sprinkle the mixture over the berries. Bake at 350 for 40 to 45 minutes. These squares freeze well if there are any leftover.

Blueberry Walnut Roll

Submitted by: Jayne Littlefield
From: Tucson, AZ
Bake Time: 25 minutes
Bake Temperature: 350 degrees

Ingredients:

1 1/3 Cups ground walnuts
1 Tsp. baking powder
1/2 Tsp. cinnamon
1 1/2 Cups blueberries

6 eggs separated
3/4 Cup sugar
3/4 Cup heavy cream
1/4 Cup confectioners' sugar

1 Tbls. rum extract
1/2 Tsp. vanilla

Instructions:

Combine walnuts, baking powder, and cinnamon - set aside. Separate 6 eggs and keep at room temperature. Preheat oven to 350, grease jellyroll pan, line with wax paper, grease and flour paper. In large mixer bowl, beat egg yokes at high speed for 1 minute. Gradually add 1/2 cup sugar, continue beating five minutes longer, until mixture is thick and pale. With a rubber spatula, fold in walnut mixture until well combined. In another large mixing bowl, beat egg whites until foamy, gradually add 1/4 cup sugar beating until stiff but not dry. Gently fold in 1/3 of the whites into the yolk mixture, and then fold in remaining whites. Pour into prepared pan, even with spatula. Bake 20-25 minutes until toothpick inserted and comes out clean. While cake is baking, generously sprinkle a clean kitchen towel with confectioners' sugar. When cake is done, immediately invert onto towel and carefully remove wax paper. Lifting from the short side, roll cake up with towel jellyroll fashion. Cool completely. Filling: In large mixer bowl beat 3/4 Cup heavy cream until soft peaks form. Add 1/4 cup confectioners' sugar, 1 tbsp. rum extract, and 1/2 tsp. vanilla extract. Beat until stiff. Fold in 1-1/2 cup blueberries. When cake is cool, gently unroll and spread on filling; roll up again and transfer, seam side down, to serving tray. Sprinkle additional confectioners sugar across top of cake. Refrigerate. Serves 10-12.

ARKANSAS

"We live in East Central Arkansas on the banks of the Mississippi River. My husband and I enjoy going there to watch the barges going up and down the river. We live on a small farm and raise soy beans, wheat and milo. Here's your recipe and I hope that your family and friends enjoy it as much as we do."

"Blueberries are low in calories, containing only 42 calories per ½ cup serving. Blueberries require no peeling, pitting, or coring. Simply wash and enjoy. Of all the berries, blueberries rank first in Vitamin A and second in food energy. They are a good source of Vitamin C and iron and supply trace minerals such as calcium, magnesium, phosphorus and potassium."

Blueberry Torte

Submitted by: Vivian Harris
From: Jonesboro, AR
Bake Time: 25 minutes
Bake Temperature: 300 degrees

Ingredients:
1 8 Oz. package cream cheese   1 Tsp. vanilla   2 Eggs
1 Cup sugar   1 graham cracker crust
1 can of blueberry pie filling of fresh blueberries

Instructions:
Mix the cream cheese, the eggs, the sugar and the vanilla with a mixer until smooth. Pour mixture into graham cracker crust. Bake 25 minutes at 300 degrees. Let cool completely. Spread 1 can of blueberry pie filling or top pie with fresh blueberries.

Cold Blueberry Soup

Submitted by: Sandy Hatcher
From: Little Rock, AR

Ingredients:

2 Cups blueberries
4 Tbls. sour cream or yogurt
1/2 Tsp. vanilla extract
1 Cup unsweetened pineapple juice
1 Tsp. fresh lemon juice

Instructions:

Put the blueberries, pineapple juice, and vanilla extract into a blender. Blend until smooth. Serve in 4 chilled bowls with a dollop of topping slightly swirled in. Sugar may be added if the blueberries are tart.

Blueberry Marmalade

Submitted by: Judy Spinks
From: Drasco, AR

Ingredients:

1 medium orange
3/4 Cup water
3 Cups crushed blueberries
1/2 of a 6 Oz. bottle of liquid fruit pectin
1 lemon 5 Cups sugar

Instructions:

Remove peel from the orange and the lemon. Scrape excess white from the peel and cut peel into very fine shreds. Place in a large saucepan and add the 3/4 cup water. Bring to a boil and simmer covered for 10 minutes stirring occasionally. Remove and discard the white membrane from the lemon and orange inner fruit. Finely chop pulp (discard seeds). Add to peel the 3 cups of blueberries. Cover and simmer 12 minutes. Add 5 cups sugar and bring to a full rolling boil. Boil hard one minute stirring constantly. Remove from heat immediately and stir in the 1/2 of the 6-ounce bottle of liquid fruit pectin. Skim off foam. Stir and skim for about 7 minutes. Ladle into hot scalded jars and seal at once. Makes 6 1/2 pints.

Blueberry Bell Crunch

Submitted by: Terry Fisher
From: Hot Springs, AR
Bake Time: 40 minutes
Bake Temperature: 350 degrees

Ingredients:

2 Cups blueberries   2 Tablespoons lemon juice   1 Cup quick oatmeal
2 Tablespoons flour   1 Cup flour   1/2 Teaspoon vanilla
3/4 Teaspoon salt   1/2 Cup brown sugar   1/2 Cup butter
1/2 Cup sugar

Instructions:

Combine the blueberries, 2 tablespoons of flour, 1/4 teaspoon salt, 1/2 cup sugar and the 2 tablespoons of lemon juice. Place in a well greased pie plate. Combine the 1 cup flour, 1 cup quick oatmeal, 1/2 cup brown sugar, 1/2 teaspoon salt, 1/2 teaspoon vanilla and 1/2 cup butter. Pack over bottom layer and bake at 350 for 30 to 40 minutes. Chill and serve with a whipped topping. Also can be served warm.

Blueberry Sour Cream Dessert

Submitted by: Winsel Harper
From: Lexa, Ar
Bake Time: 45 minutes
Bake Temperature: 350 degrees

Ingredients:

1 package white cake mix   1 Tsp. Nutmeg   1/2 Cup melted butter
1/2 Cup graham cracker crumbs   1/2 Cup chopped nuts   1 egg
1/2 Cup sour cream   (1) 21 Oz. can of pie filling or fresh blueberries

Instructions:

Heat oven to 350. In a large bowl, combine cake mix, nuts, graham cracker crumbs, and butter at low speed until crumbly. Reserve 1/2 cup for topping, but press the rest of the mixture in a 9 inch square pan. Blend sour cream, egg and nutmeg. Pour over pie filling and sprinkle with reserved crumbs. Bake at 350 for 40 to 45 minutes or until topping is golden. Serve warm or cold.

Blueberry Buckle

Submitted by: Vernita Reed
From: Benton, AR
Bake Time: 40 minutes
Bake Temperature: 375 degrees

Ingredients:

4 Tbls. butter   2 1/3 Cups flour   1/2 Cup milk
1/2 Tsp. cinnamon   1 1/4 Cups sugar   2 Tsp. baking powder
1/4 Cup butter   1 egg   1/2 Tsp. salt
2 Cups blueberries

Instructions:

Cream together 4 tablespoons butter and 3/4 cup sugar. Add egg and beat well. Sift together 2 cups flour, 2 teaspoons baking powder and 1/2 teaspoon salt. Add alternately with the milk. Beat until smooth and fold in berries. Spread batter into a 9 by 9 greased pan or a pan lined with parchment paper. For topping combine 1/2 cup sugar, 1/3 cup flour, 1/2 teaspoon cinnamon and 1/4 cup soft butter and mix until crumbly. Sprinkle over batter and bake at 375 for 35 to 40 minutes.

Double-Good Blueberry Pie

Submitted by: Norma Hannah
From: Van Buren, AR

Ingredients:

3/4 Cup sugar   1/8 Tsp. Salt   4 Cups blueberries
1 Tbls. lemon juice   3 Tbls. cornstarch   1/4 Cup water
1 Tbls. Butter   1 baked 9 inch pie shell

Instructions:

Combine sugar, cornstarch, and salt in a saucepan. Add water and 2 cups blueberries. Cook over medium heat stirring constantly until mixture comes to a boil and is thickened and clear. Mixture will be very thick. Remove from heat and stir in butter and lemon juice. Cool. Place remaining 2 cups blueberries in pie shell. Top with cooked berry mixture. Chill. Serve garnished with whipped cream or cool whip.

Fresh Blueberry Pie

Submitted by: Mildred Morgan
From: Decatur, AR

Ingredients:
1 baked 9 inch pie shell
5 Tbls. flour   4 Cups blueberries   pinch of salt
1 3/4 Cups sugar   1/4 Cup cold water   1/2 cup water

Instructions:
Make a smooth paste out of the cold water, flour, and the salt. Boil 1 cup of the blueberries with sugar and 1/2 cup water. Add the flour paste and stir until it thickens. Remove from stove and cool. When cool add the remaining blueberries and put mixture into pie shell. Refrigerate. When cold, garnish with a sweetened whipped cream or whipped topping.

Peach-Blueberry Cobbler

Submitted by: Myra Docterman
From: Horseshoe Bend, AR
Bake Time: 30 minutes
Bake Temperature: 350 degrees

Ingredients:
1 Tbls. cornstarch   1/4 Cup brown sugar   1 Cup fresh blueberries
1 Cup flour   1/2 Cup cold water   1 Tbls. butter
1/2 Cup white sugar   1/4 Cup softened butter   1 Tbls. lemon juice
1 1/2 Tsp. baking powder   2 Tbls. white sugar   1/2 Cup milk
1/4 Tsp. nutmeg   2 Cups sugared sliced fresh peaches

Instructions:
Mix the cornstarch, brown sugar, and water. Add fruits. Cook and stir until mixture thickens. Add 1 tablespoon of butter and lemon juice; stir in. Pour into an 8" round dish. In mixing bowl, add dry ingredients. Add milk and butter all at once, and beat smooth. Pour over fruit. Sprinkle 2 Tbsp. sugar and nutmeg over the batter. Bake at 350 for 30 minutes. For canned or frozen fruits, drain and use 1/2 cup fruit syrup for water.

Blueberry Coffee Cake

Submitted by: Lavina Gingerich
From: Mnt. View, AR
Bake Time: 25 minutes
Bake Temperature: 425 degrees

Ingredients:

1 1/2 Cup flour
1/2 Cup sugar
1 Tbls. baking powder
1/2 Tsp. salt

1 egg
1/2 Cup butter (divided)
1 Tbls. flour
1/2 Cup milk

3/4 Cup brown sugar
1/2 Cup chopped nuts
1 1/2 Cup blueberries

Instructions:

Mix the 1/2 cup flour, white sugar, baking powder, salt, egg, milk and 1/4 cup butter. Fold in blueberries and pour into a greased 8 by 8 inch pan. Sprinkle topping (1/4 cup butter, 1 tablespoon flour, 3/4 cup brown sugar, and 1/2 cup chopped nuts. Bake at 425 for 20 to 25 minutes.

Blueberry Buckle Coffee Cake

Submitted by: Earl Sewell
From: Vilonia, AR
Bake Time: 50 minutes
Bake Temperature: 375 degrees

Ingredients:

2 1/3 Cups flour
1 1/4 Cup sugar
1/4 Cup butter
1/4 Tsp vanilla
1/2 Cup confectioners' sugar

1 egg
2 Cups blueberries
3/4 Tsp. salt
1/4 Cup shortening

2 Tsp. hot water
3/4 Cup milk
1/2 Tsp. cinnamon
2 1/2 Tsp. baking powder

Instructions:

Heat oven to 375 and grease a 9 by 9 by 2 square inch pan. Blend 2 cups flour, 3/4 cup sugar, baking powder, salt, shortening, milk and egg. Beat 30 seconds and fold in blueberries. Spread batter in pan. Create a crumb topping of 1/2 cup sugar, 1/3 cup flour, 1/2 teaspoon ground cinnamon, and 1/4 cup butter. Bake at 375 for 45 to 50 minutes. While cake is baking make a glaze of 1/2 cup powdered sugar, 1/4 teaspoon vanilla, and 1 1/2 teaspoons to 2 teaspoons of hot water. Mix all ingredients to drizzling consistency.

Blueberry Rhubarb Breakfast Sauce

Submitted by: T.L.T.
From: Little Rock, AR
Bake Time: 10 minutes

Ingredients:

6 Cups finely chopped rhubarb   (1) 3 Oz. Pkg. of Raspberry gelatin   4 Cups sugar
(1) 21 Oz. can of blueberry pie filling

Instructions:

In a saucepan, bring rhubarb and sugar to a boil. Boil for 10 minutes. Remove from heat, add pie filling, and mix well. Return to a boil. Remove from heat and stir in gelatin. Store in jars or freezer containers. Refrigerate or freeze until ready to use. Serve with pancakes, waffles, toast, or English muffins. Makes 7 cups.

Ham with Blueberries and Peaches

Submitted by: Sandra Carbonetto
From: Harrison, AR
Bake Time: 35 minutes
Bake Temperature: 350 degrees

Ingredients:

1 8 Oz. can sliced peaches   2 Tbls. butter   1 Cup blueberries
2 Tbls. brown sugar   2 Lbs. sliced ham   1 1/2 Tsp. curry powder

Instructions:

Place ham slices in a baking pan. Pour peaches with their liquid and blueberries over ham. Mix curry powder with brown sugar and sprinkle over fruit. Dot with butter and bake at 350 degrees for 35 minutes.

Peach and Blueberry Cream Parfaits

Submitted by: Sandra Carbonetto
From: Harrison, AR

Ingredients:

1/2 Cup sugar
3 Tbls. cornstarch
1 Cup sweetened sliced peaches

2 Cups milk
2 eggs, slightly beaten
1 Cup fresh blueberries

1 Tsp. vanilla
1/4 Tsp. salt

Instructions:

Mix sugar, cornstarch, and salt in a saucepan. Stir in milk. Cook over low heat, stirring constantly, until thickened. Blend small amount of hot mixture into eggs, return to cooked mixture. Cook for two minutes longer over low heat, stirring constantly. Cool and blend in vanilla. Chill. Alternate layers of chilled cream pudding, peaches, and blueberries in tall tumblers or parfait glasses. Repeat layers to fill glasses. Top with a spoonful of pudding and one peach slice. Serve cold. makes 6 servings.

CALIFORNIA

"I was a teacher for 26 years in elementary school. My town is about an hour's drive from Yosemite National Park. The Park is a wonderful part of our history, the gold rush of '49. We have a small museum here, which many children come to see. I hope that someday you may come here and see this area as well as enjoy the beauty of Yosemite National Park. I have enclosed a recipe…"

Blueberry Apple Pizza

Submitted by: Dottie Gunnarson
From: Seal Beach, CA
Bake Time: 25 minutes
Bake Temperature: 350 degrees

Ingredients:

1 pkg. wild blueberry muffin mix	1/4 Cup butter	1/4 Cup brown sugar
3/4 Cup quick oatmeal	1 egg	2 Tbls. butter
1 Cup chopped nuts	(1) 21 Oz. can apple pie filling	

Instructions:

For pizza crust grease 12 inch pizza pan. Drain berries from mix. Combine muffin mix, oatmeal, 1/2 cup chopped nuts, 1/4 cup butter, and egg. Press into a pizza pan and form ridge at edge. Bake at 350 for 10 to 15 minutes. Spread pie filling over crust and sprinkle with blueberries. In a bowl combine the rest of the nuts, brown sugar, and the rest of the butter. Sprinkle over berries and bake at 300 for 15 to 20 minutes. Cut into wedges and serve warm.

Blueberry Rice Supreme

Submitted by: Ellen Saulsbury
From: Lodi, CA

Ingredients:

1/3 Cup ice water   1/2 Cups flaked coconut   2 Tbls. lemon juice
1/3 Cup chopped walnuts   2 Tbls. sugar   1/3 Cup nonfat dry milk
1 1/2 Cup fresh blueberries   2 Cups boiled cooled rice (1 Cup uncooked)

Instructions:

Mix together the rice, coconut, walnuts, and blueberries. In a separate 1 quart bowl measure water and milk. Beat until stiff. Add sugar and lemon juice gradually, beating constantly. Chill 1/2 hour before serving. Mix together the rice mixture and the topping just before serving.

Bumbleberry Black Bottom Pie

Submitted by: Vic Van Dreser
From: Sunol, CA
Bake Time: 25 minutes
Bake Temperature: 350 degrees

Ingredients:

4 Oz. bittersweet chocolate   1/2 Tsp. cinnamon   3 Tsp sugar
2 Tbls. orange juice   Pinch of salt   Shaved chocolate
4 eggs separated   1 Cup blueberries   1/2 Cup sugar
1 Cup blackberries, 1 Cup raspberries, 1 Cup blueberries or any combination of the three
1 Cup heavy cream whipped with 2 Tbls. powdered sugar and 2 Tbls. orange or berry liqueur (optional)

Instructions:

Preheat oven to 350 degrees. Melt chocolate with orange juice in a small heavy saucepan or double boiler over low heat. Stir until smooth and let cool. Butter a 10 inch pie plate. Beat yolks with sugar until very thick and pale in color. Add cinnamon and melted chocolate, beating slow until blended. Beat whites with salt until stiff. Add whites 1/3 at a time to mixture, folding in with a spatula. Pour mixture into a pie plate, level with a spatula, and bake for about 25 minutes. Allow to cool. As crust cools, it will sink in the center, forming a shell. In a large bowl, toss berries with sugar. Fill cooled pie shell with berry mixture. Spread with whipped cream and sprinkle with shaved chocolate.

Blueberry Syrup

Submitted by: Debbie Brothwell
From: Torrance, CA

Ingredients:

5 Cups fresh blueberries   1/2 Cup water   1 Cup sugar
1 Tsp. lemon juice   2 Tbls. sugar   2 Tablespoons water

Instructions:

Combine the berries, sugar, and water in a medium-sized heavy saucepan over medium high heat and bring to a boil, stirring. Boil gently and cook until berries pop and look shriveled, about 10 minutes. Stir in lemon juice, and remove from heat. Allow the syrup to cool, then ladle into jars. Cover and refrigerate for 2-3 weeks. Serve over ice cream, pancakes, waffles, or biscuits.

Blueberry Tea Cake

Submitted by: Betty Mehlenbacher
From: Chula Vista, CA
Bake Time: 35 minutes
Bake Temperature: 350 degrees

Ingredients:

2 Tbls. butter   1 1/2 Cups flour   1 Cup sugar
1/3 Cup milk   2 eggs, separated   1 1/2 Cup blueberries

Instructions:

Cream the butter and sugar together until well blended. Beat egg yolks and add the flour alternately with the milk. Fold in the stiffly beaten egg whites. Pour half the batter into a greased 13 by 9 baking pan. Cover with the floured berries and then with the remaining batter. Bake in a oven at 350 for 35 minutes. Sprinkle with powdered sugar while still warm and cut into squares.

French Toast Bake

Submitted by: Lisa Benton
From: La Mesa, CA
Bake Time: 35 minutes
Bake Temperature: 400 degrees

Ingredients:

1 Cup packed brown sugar   1 Cup chopped pecans   5 eggs
1 Tsp. vanilla extract   1/4 Cup butter melted   2 1/2 Cups milk
1/2 Tsp. ground nutmeg   2 Cups blueberries
12 slices day old French bread (1 inch thick)

Instructions:

Arrange the bread in a greased 13 by 9 inch baking dish. In a bowl, combine the eggs, milk, 3/4 cup brown sugar, vanilla, and nutmeg. Pour over bread. Cover and refrigerate for 8 hours or overnight. Remove from the refrigerator 30 minutes before baking. Sprinkle pecans over egg mixture. Combine butter and remaining sugar; drizzle over the top. Bake uncovered at 400 degrees for 25 minutes. Sprinkle with blueberries. Bake 10 minutes longer on until a knife inserted near the center comes out clean. Makes 6-8 servings.

No-Dough Blueberry-Peach Cobbler

Submitted by: Esther Leiggi
From: Whittier, CA
Bake Time: 55 minutes
Bake Temperature: 350 degrees

Ingredients:

1/2 Cup butter   2 Tsp. baking powder   1/2 Cup sugar
1 Cup flour   1/2 Cup milk   2 Cups blueberries
3/4 Cup sugar   2 Cups freshly sliced peaches

Instructions:

Melt butter in a 2-1/2 quart baking dish. Set aside. Combine flour, 3/4 cup sugar, and baking powder; add milk, and stir until blended. Pour batter over butter in baking dish. Do not stir. Combine peaches, blueberries, and 1/2 cup sugar; spoon over batter. Do not stir. Bake at 350 degrees for 45 to 55 minutes. Makes 6 servings.

Double Blueberry Pancakes

Submitted by: Ellen Saulsbury
From: Lodi, CA

Ingredients:

2 1/2 Cups Original Bisquick Mix   1/3 Cup sour cream or plain yogurt   1 1/4 Cups milk
2 eggs   1/3 Cup sugar   2 Cups blueberries

Instructions:

Stir Bisquick, milk, sugar, sour cream, and eggs in a large bowl until blended. Gently stir in blueberries. Pour by slightly less than 1/4 cupful onto hot griddle (grease as necessary). Cook about two minutes or until pancakes are dry around edges. Turn, cook about 2 minutes, or until golden. Makes about 23 pancakes.

Blueberry Cherry Pie

Submitted by: Susan Pearce
From: Anderson, CA
Bake Time: 55 minutes
Bake Temperature: 425 degrees

Ingredients:

2 Tbls. quick-cooking tapioca   1/2 Cup cherry juice   1/2 Cup sugar
1/2 Tbls. lemon juice   1/4 Tsp. salt   2 Cups blueberries
Pastry for 2 crust 9 inch pie   1 Tbls. butter
2 Cups drained pitted canned red sour cherries

Instructions:

Combine tapioca, sugar, salt, blueberries, cherries, cherry juice and lemon juice. Roll half of the pastry 1/8 inch thick into 9 inch pie pan and turn edges. Pour filling into pie shell. Dot with butter. Moisten edge of bottom crust. Fit on top rolled out crust over filling and fit and pinch edges. Slit top for venting. Bake at 425 for 55 minutes or until bubbly.

Blueberry Yogurt Dessert

Submitted by: Susan Pearce
From: Anderson, CA

Ingredients:

1/2 Cup blueberry yogurt   2 Tbls. sugar   1 Quart fresh blueberries

Instructions:

In a small bowl, combine yogurt and sugar. Divide blueberries between four individual serving dishes. Drizzle with yogurt mixture. Makes 4 servings.

Blueberry Chills

Submitted by: Betty Boyle
From: Escondido, CA

Ingredients:

24 Crushed vanilla wafers   1 Cup whipping cream   1 Can blueberry pie filling
1/2 Cup confectioners' sugar   8 Oz. pkg. of cream cheese

Instructions:

Spread wafers in a 8 inch pan. Beat cream cheese, sugar, and whipping cream until smooth. Place over crumbs. Top with a can of blueberry pie filling and then freeze for at least 2 hours.

Blueberry Cream Cheese Pie

Submitted by: Betty Dorsey
From: Mariposa, CA

Ingredients:

1 13 Oz. pkg. cream cheese   1 Can blueberry pie filling   1/2 Cup sugar
1/2 Pint cream   (1) 9 inch baking pie shell   1 Tsp. vanilla

Instructions:

Soften cream cheese to room temperature. Whip cream, sugar, and vanilla. When thickened add cream cheese. Pour into a pie shell and chill overnight. Add blueberries 1-2 hours before serving.

Blueberry Cake

Submitted by: Pauline Trabucco
From: Mariposa, CA
Bake Time: 40 minutes
Bake Temperature: 350 degrees

Ingredients:

1 Cup blueberries   1 Cup sugar   2 Tsp. baking powder
1/3 Cup sugar   1/4 Cup flour   1 egg, beaten
1/2 Tsp. Salt   1 Tsp. cinnamon   1/4 Cup butter
1 3/4 flour   1/2 Cup milk

Instructions:

Mix blueberries with 1/4 cup of the flour. Cream the sugar and butter well and add the beaten egg. Sift the flour with baking powder and salt and add alternately with the milk. Beat well after each addition. Fold in blueberries. Pour into buttered 8 by 8 inch square pan and top with a mixture of sugar and cinnamon. Bake at 350 for 40 minutes. to make cupcakes, prepare same batter and bake in a buttered cupcake pan for 25 minutes.

French Blueberry Glace Pie

Submitted by: Helen Whitechat
From: Red Bluff, CA

Ingredients:

1 Quart. Blueberries	3 Tbls. cornstarch	1 Cup water
(1) 3 Oz. pkg. of cream cheese	1 Cup sugar	(1) pre-baked piecrust

Instructions:

Wash and drain blueberries. Simmer 1 cup blueberries and 2/3 cup water about 3 minutes. Blend sugar, cornstarch, and remaining 1/3 cup water. Add to boiling mixture. Boil one minute stirring constantly. Cool. Spread cream cheese over bottom of a cooled pie shell. Save out 1/2 cup choice berries: put remaining 2/12 cups berries in baked pie shell. Cover with cooked mixture and garnish with the 1/2 cup berries. Refrigerate until firm- about 2 hours. Serve with sweetened whipped cream and/or ice cream.

Bonnie's Blueberry Dessert

Submitted by: Mrs. Edwin Bossen
From: Desert Hot Springs, CA
Bake Time: 16 minutes
Bake Temperature: 350 degrees

Ingredients:

20 graham cracker squares	1 Can sweetened condensed milk	1/3 Cup melted butter
1/2 Cup fresh lemon juice	8 Oz. Pkg. cream cheese	8 Cups blueberries

Instructions:

Crush the 20 crackers into a bowl. Stir in the 1/3 cup butter (melted). Pat down mixture into a 10 by 14 inch jelly roll pan to form a crust. Bake 16 minutes in a 350 degree oven. Cool. Beat the cream cheese (room temperature), condensed milk and lemon juice until smooth. Fold in blueberries(preferably large because they look so wonderful). Spread carefully onto cooled crust. Refrigerate overnight or up to 24 hours before serving.

Blueberry Brunch Cake

Submitted by: Kathleen Bergstrom
From: North Hollywood, CA
Bake Time: 40 minutes
Bake Temperature: 325 degrees

Ingredients:

1 1/4 Cup flour
1/3 Cup sugar
1 1/2 Tsp. baking powder
1 egg, beaten

1/3 Cup salad oil
1/2 Cup milk
1 Cup blueberries
1 Tbls. lemon juice

1/3 Cup brown sugar
1/2 Cup chopped nuts
1/4 Tsp. cinnamon
1 Tbls. butter

Instructions:

Combine one cup flour, the white sugar, and the baking powder. Mix well and add egg, milk, salad oil, and lemon juice. Stir until blended; pour into greased 8 inch square pan. Scatter blueberries over the top. Create a topping of the brown sugar, flour, cinnamon, nuts and butter. Sprinkle over the top and bake at 325 for 40 minutes.

Triple Fruit Pie

Submitted by: Mrs. William Tade
From: Carlsbad, CA
Bake Time: 400 degrees
Bake Temperature: 50 minutes

Ingredients:

1/2 Tsp. almond extract
1/4 Tsp. ground nutmeg
1 1/4 Cup each of fresh blueberries, raspberries and chopped rhubarb
pastry for a double 9 inch pie crust

1 1/4 Cups sugar
1/4 Cup quick cooking tapioca

1/4 Tsp. salt
1 Tbls. lemon juice

Instructions:

In a large bowl, combine fruits and extract; toss to coat. In another bowl, combine sugar, tapioca, nutmeg, and salt. Add to fruit; stir gently. Let stand for 15 minutes. Line a 9 inch pie plate with bottom crust; trim pastry even with edge. Stir lemon juice into fruit mixture; spoon into the crust. Roll out remaining pastry; make a lattice crust. Seal and flute edges. Bake at 400 degrees for 20 minutes. Reduce heat to 350 and bake 30 minutes longer or until the crust is golden brown and the filling is bubbly. Creates 6-8 servings. Frozen berries and rhubarb may be substituted for fresh. Thaw and drain before using.

Blueberry- Lemon Cream Pie

Submitted by: A friend
From: Arleta, CA

Ingredients:

1 Cup sugar
3 Tbls. cornstarch
1 Tbls. shredded lemon peel
1) 8 Oz. carton sour cream

3 beaten egg yolks
1/4 Cup butter
2 Cups fresh blueberries

(1) 9 inch pie shell
(1 Cup milk
1/4 Cup lemon juice

Instructions:

In a saucepan combine 1 cup sugar and cornstarch. Add milk, egg yolks, butter, and 1 tablespoon lemon peel. Cook and stir over medium heat until thickened and bubbly; cook and stir 2 minutes more. Remove from heat; stir in lemon juice. Transfer to a bowl; cover surface with plastic wrap and refrigerate until cool. When cool, stir in sour cream and blueberries into mixture; pour into pastry shell. Cover and chill at least 4 hours. If desired, stir a little lemon peel into sweetened whipped cream. Spoon atop pie and garnish with lemon slices. Creates 8 servings.

Blueberry Lemon Muffins

Submitted by: Debra Jones
From: Los Gatos, CA
Bake Time: 20 minutes
Bake Temperature: 375 degrees

Ingredients:

1 Cup blueberries
2 Cups flour
3/4 Cup sugar

2 Tsp. baking powder
1/4 Tsp. Salt
1 egg

1 Tbls. grated lemon peel
1/2 Cup milk
1/4 Cup salad oil

Instructions:

Rinse and drain blueberries. In a bowl combine flour, sugar, baking powder, and salt. Beat egg to blend with oil, then add milk and lemon peel and pour into flour mixture. Stir just until the dry ingredients are evenly moistened. Batter will be stiff. Gently mix in the blueberries. Spoon batter evenly into greased or paper lined 2 or 2 1/2-inch muffin cups. Bake in 375 degrees until golden, about 20 minutes. Turn muffins on their sides in pan to cool slightly. Serve warm or cool. Makes 12 muffins and freezes wonderfully.

Blueberry Oatmeal Cookies

Submitted by: St. Catherine Labarre Church / Global Publishing Co.
From: San Diego, CA
Bake Time: 10 minutes
Bake Temperature: 375 degrees

Ingredients:

1 pkg. blueberry muffin mix   1/3 Cup cooking oil   1 Tbls. milk
1/4 Cup brown sugar   3/4 Cup quick cooking oats   1 egg

Instructions:

Preheat oven to 375. Drain the blueberries out of the muffin mix can. Place on a paper towel. In a bowl combine all ingredients. Mix well. Drop from a teaspoon onto an ungreased cookie sheet. Make a deep depression in each cookie and fill with 7 to 8 well drained blueberries. Push dough from sides to cover berries and pat down. Bake at 375 for 8 to 10 minutes.

Lemon Pudding with Blueberry Sauce

Submitted by: Anonymous
From: Santa Barbara, CA
Bake Time: 30 minutes
Bake Temperature: 325 degrees

Ingredients:

2 Tbls. fresh lemon juice   1/4 Cup sugar   1 Tsp. butter
1/3 Cup 2% low fat milk   1 Tbls. sugar   1 Cup fresh blueberries
1 Tsp. cornstarch   1 Tsp. grated lemon rind 1 Tsp. fresh lemon juice
1/4 Cup thawed egg substitute   2 Tbls. flour   1 Tbs. water

Instructions:

Beat egg substitute at high speed of an electric mixer for 1 minute. Add 1/4 cup sugar; beat 1 minute. Add milk, 1/2 teaspoon grated lemon rind, two tablespoons lemon juice, and the butter (melted). Beat well. Add flour; beast until well blended. Pour half the mixture into each of 2 (6 ounce) custard cups that are coated with cooking spray. Place custard cups in a 9 inch square baking pan. Add water to pan to a depth of 1 inch. Bake at 325 for 30 minutes or until set. Remove cups from water and set aside.

Combine 1 tablespoon sugar, water, cornstarch, 1/2 teaspoon lemon rind, and 1 teaspoon of lemon juice into a microwave-safe bowl; stir well. Add blueberries; stir well. Microwave at high for 3 minutes of until thickened and bubbly, stirring after 1 1/2 minutes. Spoon half of the mixture over each pudding. Creates 2 servings.

Blueberry Citrus Cake

Submitted by: Kathy Hastings
From: Mission Viejo, CA
Bake Time: 40 minutes
Bake Temperature: 325 degrees

Ingredients:

1 pkg. lemon cake mix   3 eggs   1/2 Cup orange juice
1 1/2 Cups blueberries   1 Tbls. finely grated lemon peel   1/2 Cup water
1 Tbls. finely grated orange peel   1/3 Cup vegetable oil

Instructions:

Preheat oven to 325. Grease and lightly flour two 8 or 9 inch round cake pans. In a large mixing bowl combine cake mix, orange juice, water, oil and eggs. Beat with an electric mixer on low for 30 seconds and then on medium for 2 minutes. Fold in blueberries, orange, and lemon peel. Pour into prepared pans and bake for 35 to 40 minutes. Cool in pans on a rack for 10 minutes. Remove cakes from pans and thoroughly cool on racks before frosting with Citrus Frosting.

Citrus Frosting: In a medium bowl beat together a 3 ounce package of softened cream cheese, 1/4 cup softened butter. Beat until fluffy. Add 3 cups powdered sugar and 2 tablespoons orange juice. Beat until combined. In a small bowl beat 1 cup whipping cream to soft peaks; add to cream cheese mixture. Add 2 tablespoons finely grated orange peel and 1 Tablespoon finely grated lemon peel. Beat on low until combined.

COLORADO

"I remember as a boy, picking wild blueberries in the woods on the Massachusetts coast, although I suspect they may have been planted by some early settlers long ago…"

"I love blueberries and do not understand people who don't. But over the years when I have served this recipe, I have been told several times, " For the first time, I liked and enjoyed blueberries." Good luck with your recipe collection."

Quick Blueberry Cobbler

Submitted by: Alice Allen
From: Boulder, CO
Bake Time: 30 minutes
Bake Temperature: 350 degrees

Ingredients:

3/4 Cup butter	2 Tsp. baking powder	1/2 Cup sugar
1 Cup flour	3/4 Cup milk	3/4 Cup sugar
2 Cups blueberries		

Instructions:

Melt the butter into a 7 by 11 inch oblong pan. Preheat oven to 350 degrees. Mix together 1 cup flour, 3/4 cup sugar, the baking powder, and milk. Pour this into melted butter. Add the blueberries. Sprinkle with 1/2 cup sugar on top. Bake 25 to 30 minutes at 350 degrees. Enjoy with or without ice cream.

Cornmeal Blueberry Muffins

Submitted by: Betty Sloan
From: Denver, CO
Bake Time: 20 minutes
Bake Temperature: 400 degrees

Ingredients:

1 Cup flour	2 Tsp. baking powder	1 Cup buttermilk
2/3 Cup cornmeal	1/2 Tsp. baking soda	1/4 Cup melted butter
1/3 Cup sugar	1/2 Tsp. salt	1 1/3 Cup blueberries
3 eggs		

Instructions:

Combine the flour, cornmeal, sugar, baking powder, baking soda, and salt. In a separate bowl mix together the beaten eggs, buttermilk, and melted butter. Combine the two mixtures until just moist. Fold in the blueberries and spoon into sprayed muffin pans. Bake at 400 degrees for 20 minutes. Creates about a dozen muffins.

Blueberry Oat Cake

Submitted by: Carmela Zarlengo
From: Denver, CO
Bake Time: 50 minutes
Bake Temperature: 375 degrees

Ingredients:

2 eggs	1 Tsp. baking soda	2 Cups quick cooking oats
2 Cups buttermilk	2 Cups flour	1 Tsp. ground cinnamon
1 Cup brown sugar	2 Tsp. baking powder	1/2 Cup vegetable oil
1/2 Tsp. salt	1 Cup chopped nuts-optional	2 Cups blueberries

Instructions:

In a mixing bowl, beat the eggs, buttermilk, brown sugar, and oil. Combine the flour, baking powder, baking soda, cinnamon, and salt; add to the batter. Beat on low speed for 2 minutes. Fold in oats; blueberries and if desired walnuts. Transfer to a greased and floured 10 inch fluted tube pan. Bake at 375 for 45 to 50 minutes or until a toothpick comes out clean. Cool for 10 minutes before removing from pan to a wire rack to cool completely. Dust with confectioners' sugar. Creates 12 to 16 servings.

Blueberry Cottage Danish Snack

Submitted by: Lucille Creed
From: Englewood, CO

Ingredients:

1 slice of bread   1 sprinkle of sugar   1 handful of blueberries
1 spoonful of cottage cheese

Instructions:

For a wonderful simple snack, slice of a piece of bread and cover with cottage cheese. Place blueberries over the entire top. Sprinkle with sugar and microwave a few seconds until the topping is bubbly.

Blueberry Custard

Submitted by: Betty
From: Colorado Springs, CO
Bake Time: 50 minutes
Bake Temperature: 350 degrees

Ingredients:

4 slices of white bread   1 Tsp. ground cinnamon   1 Tsp. vanilla
2 Tbls. Butter, softened   2 Cups milk   1 1/2 Cups blueberries
1/2 Cup sugar   3 eggs, beaten

Instructions:

Trim crusts from bread and spread one side with butter. Cut each slice into four squares. Arrange in lightly buttered 8 inch square baking dish with butter side up. Sprinkle with blueberries and cinnamon. Heat milk and sugar until warm; do not boil. Stir until sugar dissolves. Combine warm milk mixture, eggs, and vanilla. Pour over blueberries. Set dish in a larger pan then add hot water to a depth of 1 inch. Bake at 350 degrees for 45 to 60 minutes or until toothpick comes out clean. Remove custard from water and cool 10 minutes or chill before serving. Creates 6 servings.

Blueberry Bread

Submitted by: Patty Barrow
From: Bristol, CO
Bake Time: 60 minutes
Bake Temperature: 375 degrees

Ingredients:

1 1/2 Cups flour	1 Cup sugar	1 1/2 Cup blueberries
1 Tsp. baking powder	1/3 Cup milk	1 Tsp. vanilla
1/2 Cup butter	2 eggs	

Instructions:

Cream together butter, sugar, and eggs. Mix flour and baking powder. Mix dry ingredients with creamed mix alternating milk. Add vanilla and blueberries; lightly butter top of bread. Pour into greased and floured loaf pan. Bake for 1 hour at 375 degrees.

Blueberry Smoothie

Submitted by: Abbie Larsen
From: Longmont, CO

Ingredients:

1 Cup orange juice	4 Tbls. blueberry yogurt	1 dash of nutmeg
1 Cup vanilla soy milk	2 Tbls. wheat germ	2 Cups frozen blueberries
1/2 of a banana		

Instructions:

Add ingredients to blender and mix for 30 seconds or until blueberries are well blended. Puree for 10 seconds. Pour into tall glasses and enjoy.

Blueberry Loaf

Submitted by: Pearl Williams
From: Westminster, CO
Bake Time: 50 minutes
Bake Temperature: 350 degrees

Ingredients:

1 Cup flour	2 Tsp. baking powder	1 Tbls. vegetable oil
1/3 Cup brown sugar	1/2 Tsp. baking soda	3/4 Cup orange juice
1/2 Tsp. salt	1 egg	1 Cup fresh blueberries
3/4 Cup whole wheat flour	1/2 Cup shreds of wheat bran cereal	

Instructions:

Combine flours, cereal, brown sugar, baking powder, soda, and salt in a medium bowl, stirring until well combined. Set aside. Combine orange juice, egg, and oil in a large bowl; beat at medium speed of an electric mixer until well blended. Gradually add flour mixture, stirring just until moistened. Gently fold in blueberries. Spoon batter into a 8 1/2 by 4 1/2 by 3 inch loaf pan coated with cooking spray. Bake at 350 degrees for 50 minutes or until toothpick comes out clean. Cool and then slice. Yields 16 slices.

Blueberry Peach Pound Cake

Submitted by: H. Marta
From: Colorado Springs, CO
Bake Time: 70 minutes
Bake Temperature: 350 degrees

Ingredients:

1/2 Cup butter softened	2 Cup blueberries	1 1/4 Cups sugar
1/4 Cup milk	3 eggs	2 Tsp. baking powder
2 1/2 Cups flour	2 1/4 Cups chopped fresh peaches	

Instructions:

In a mixing bowl, cream butter and sugar. Beat in eggs, one at a time. Beat in milk. Combine the flour and baking powder; add to creamed mixture. Stir in peaches and blueberries. Pour into a greased and floured 10 inch fluted tube pan. Bake at 350 for 60 to 70 minutes or until a toothpick comes out clean. Cool in pan for 15 minutes; remove to a wire rack to cool completely. Dust with confectioners' sugar if desired. Creates 10 to 12 servings.

Blueberry Stuffed French Toast

Submitted by: Anthony Plezia
From: Colorado Springs, CO
Bake Time: 60 minutes
Bake Temperature: 350 degrees

Ingredients:

12 Slices French bread   12 large eggs   1/3 Cup maple syrup
1 Cup water   2 8 Oz. pkg. Cream cheese   2 Cups fresh blueberries
2 Cups milk   1 Cup sugar   1 Tbls. butter
2 Tbls. cornstarch or arrowroot

Instructions:

Grease a 13 by 9 inch baking pan. Slice the bread into 1 inch cubes. Cut the cream cheese also into 1 inch cubes. Place half of the bread crumbs evenly in prepared pan. Scatter cream cheese over bread and sprinkle with 1 cup blueberries. Arrange remaining bread crumbs over blueberries. In a large bowl combine eggs, maple syrup and milk and wisk to blend. Pour evenly over reserved bread mixture. Cover with foil and chill overnight. Preheat oven to 350 degrees. Bake, covered with foil, in the middle of the oven for 30 minutes. Remove foil and continue baking 30 more minutes, or until golden brown. In a small sauce pan combine sugar, cornstarch, and water over medium high heat. Cook stirring occasionally, 5 minutes until thickened. Stir in 1 cup blueberries and simmer, stirring occasionally, 10 minutes or until most berries burst. Add butter and stir mixture until butter is melted. (May be prepared 1 day ahead. Refrigerate and reheat gently when ready to serve.) Transfer to serving bowl. Cut French toast into serving size pieces and place on serving plates. Top with blueberry syrup and serve.

Connecticut

"It is rumored that when The Pilgrims stepped onto the New England shores, they were greeted by friendly Indians that had dried blueberries in huge wicker baskets. Historical records state that blueberries were one of the foods that helped them survive that first harsh winter in the new world."

Creamy Blueberry Pie

Submitted by: Valerie Langworthy
From: Stonington, CT
Bake Time: 55 minutes
Bake Temperature: 350 degrees

Ingredients:

3 Cups fresh blueberries
1 9 inch deep dish pie crust
1 1/2 Cup sugar

1/3 Cup flour
1/8 Tsp. Salt
2 eggs

1/2 Cup flour
1/4 Cup butter
1/2 Cup sour cream

Instructions:

Combine 1 cup sugar, 1/3 cup flour, and salt. Add eggs and sour cream, stirring until blended. Place blueberries in pastry shell, and spoon sour cream mixture over berries. In another bowl, combine 1/2 cup sugar and 1/2 cup flour. Cut in butter with pastry blender until mixture resembles course meal. Sprinkle this mixture over sour cream mixture and berries in the pie shell. Bake at 350 degrees for 50 to 55 minutes, or until lightly browned. Garnish with additional blueberries.

Blueberry Torte

Submitted by: Jayne Crepeau
From: Southbury, CT
Bake Time: 30 minutes
Bake Temperature: 350 degrees

Ingredients:
1 1/4 graham cracker crumbs
1 Cup sugar
2 eggs

4 Tbls. Cornstarch
8 Oz. pkg. cream cheese

2 Cups blueberries
1/3 Cup melted butter

Instructions:
Mix the graham cracker crumbs (about 10 crackers), 1/4 cup sugar, and 1/3 cup melted butter. Pat mixture in bottom of 9-inch square pan. Mix cream cheese and 1/2 cup sugar until smooth. Add the eggs and beat well. Pour over crumbs. Bake at 350 degrees for 25 to 30 minutes. Place blueberries, cornstarch, and 1/4 cup sugar on top of a double boiler and cook until thick. Cool. Add blueberry mixture to crust and chill overnight. Cut into squares and serve with whipped cream.

Blueberry Oatmeal Muffins

Submitted by: Eleanor Kapsia
From: Glastonbury, CT
Bake Time: 20 minutes
Bake Temperature: 400 degrees

Ingredients:
3 Cups biscuit mix
1/2 Cup brown sugar
3/4 Cup quick cooking oatmeal

2 eggs, beaten
1 Tsp. cinnamon
1 1/2 Cups milk

2 Cups fresh blueberries
1/4 Cup melted butter

Instructions:
Combine biscuit mix, brown sugar, oatmeal, and cinnamon. Mix eggs, milk, and butter. Add dry ingredients all at once and stir until just blended. Fold in blueberries. Spoon into greased muffin pans filling each cup 2/3 full. Bake at 400 degrees for 15 to 20 minutes, or until golden brown. Remove from pans and place in rack to cool. Makes 18 muffins.

Blueberry Salad

Submitted by: Elaine Macke
From: Lebanon, CT

Ingredients:

(1) 6 Oz. Pkg. black cherry Jell-O   (1) 8 Oz. pkg. cream cheese   1 Cup sour cream
(1) 20 Oz. can crushed pineapple   1/2 Cup sugar   1/2 Cup chopped nuts
(1) can blueberry pie filling   1 additional Cup of Water

Instructions:

Take juice from pineapple and add enough water to make 2 cups liquid. Heat, add Jell-O. Stir until dissolved. Add another cup cold water, pie filling and pineapple. Let congeal in 11 by 7 inch pan. Mix cream cheese, sugar and sour cream. Add nuts, pecans or walnut is suggested. Spread over congealed Jell-O.

Blueberry Yum-Yum Squares

Submitted by: Mary O'Neill
From: Vernon, CT
Bake Time: 20 minutes
Bake Temperature: 350 degrees

Ingredients:

2 Cups blueberries   1/4 Cup cornstarch   1/2 Cup butter
(1) 8 Oz. pkg. cream cheese   2 Cups sugar   3 Tbl. water
(1) Cup chopped pecans   (1) 8 Oz. container Cool Whip   1/4 Cup water
1 Cup flour

Instructions:

Combine blueberries, 1 cup sugar, and 1/4 cup water in a medium saucepan; cook over low heat until the blueberries are soft, about 15 minutes. Combine cornstarch and 3 tablespoons water in a small mixing bowl. Stir well. Add cornstarch mixture to blueberries, stirring constantly until thickened. Set aside to cool. Preheat oven to 350 degrees. Combine flour, butter, and pecans in a small mixing bowl. Mix well. Press dough evenly into a 9 by 12 by 2 inch pan. Bake for 20 minutes. Cool. Combine cream cheese and 1 cup sugar, beat until smooth, fold in Cool Whip. Spread evenly over crust. Spread blueberry mixture over top. Refrigerate several hours before serving. Cut into squares.

Blueberry Buttermilk Bread

Submitted by: Myrtle Brosz
From: Danbury, CT
Bake Time: 60 minutes
Bake Temperature: 350 degrees

Ingredients:

2 Cups flour   1/2 Tsp. ground nutmeg   3/4 Cup brown sugar
2 Tsp. baking powder   3 Tbls. melted butter   1 Cup blueberries
1 Tsp. baking soda   1 egg   3/4 Cup buttermilk
1 Tsp. salt

Instructions:

Mix flour, baking powder, baking soda, salt, and nutmeg in a medium bowl. Beat butter and sugar until well blended. Beat in egg. Beat in flour mixture alternately with buttermilk. Batter will be stiff. Stir in blueberries. Turn batter into a greased loaf pan (8 1/2 by 4 1/2 by 2 1/2 inch) Bake in 350 degree oven for 50 to 60 minutes. Cool 15 minutes and remove.

Blueberry Crunch

Submitted by: Lillian Torizzo
From: Torrington, CT
Bake Time: 40 minutes
Bake Temperature: 350 degrees

Ingredients:

1 can crushed pineapple   1/4 Cup sugar   1/2 Cup butter
3 Cups blueberries   1 Cup chopped pecans
1 box yellow cake mix   3/4 Cup sugar

Instructions:

Grease 9 by 13 pan inch pan. Spread undrained pineapple, then add a layer of blueberries and 3/4 cup sugar. Sprinkle dry cake mix over this. Drizzle melted butter all over on top with pecans and 1/4 cup sugar. Bake at 350 degrees for 35 to 40 minutes. After about 25 minutes cut holes in top to let juice come through mix. Serve with ice cream.

Blueberry Dumplings

Submitted by: Gloria Reynolds
From: Stafford Springs, CT

Ingredients:

2 1/2 Cup blueberries
1/3 Cup sugar
1 Tbls. lemon juice

1 Tbls. butter
1 Cup flour
2 Tsp. baking powder

1 pinch of salt
1/2 Cup milk

Instructions:

Combine blueberries, 1/3 cup sugar, dash of salt, and 1 cup water in a saucepan. Bring to a boiling point. Reduce heat; cover and simmer for 5 minutes. Stir in salt and 2 tablespoons more sugar together. Cut in butter until coarse meal consistency. Add milk all at once; stir only until flour mixture is dampened. Drop from tablespoon into simmering blueberry mixture. Cover tightly: cook over low heat for 10 minutes. Serve with cream. Makes 6 servings.

Blueberry Muffins

Submitted by: Elizabeth Peters
From: Danbury, CT
Bake Time: 15 minutes
Bake Temperature: 425 degrees

Ingredients:

1 Cup white flour
1 Cup whole wheat flour
1/2 Cup sugar
1 Tbls. baking powder

1/2 Tsp. Salt
1/2 Tsp. Cinnamon
1/2 Cup milk
1/2 Cup butter

2 eggs
1/2 Tsp. vanilla
1 1/2 Cups blueberries

Instructions:

Combine dry ingredients. Melt butter; add milk, eggs, and vanilla to dry ingredients. Fold in blueberries. Bake in muffin cups at 425 degrees for 15 minutes. Cool in pan for 5 minutes before taking muffins out.

Peek-A-Blue berry farm

DELAWARE

"My husband is the primary cook in our family. He makes blueberry pancakes for guests and especially for the grandchildren. Our grandson would eat them and have blueberry all over his face- we nicknamed him Blueberry Man and have great pictures of him that way. The tradition in our household was for each child to choose what they wanted for their birthday because none of them like birthday cake itself. Our oldest son always wanted blueberry pie…"

Lemon Blueberry Pancakes

Submitted by: Janice Atchley
From: Wilmington, DE

Ingredients:

1 egg
1 Cup flour
3/4 Cup milk
1 Tsp. lemon juice

1/2 Tsp. Salt
2 Tbls. vegetable oil
1 Tbls. baking powder

2 Tsp. grated lemon rind
1/2 Cup blueberries
1 Tbls. sugar

Instructions:

Beat egg until fluffy; beat in remaining ingredients except blueberries just until smooth. Stir in blueberries. Grease heated griddle. For each pancake, pour about 3 tablespoons of batter from a large spoon or from a pitcher onto a hot griddle. Cook pancakes until puffed and dry around the edges. Turn and cook.

Diabetic Blueberry Muffins

Submitted by: Ruby Parker
From: Hartly, DE
Bake Time: 20 minutes
Bake Temperature: 375 degrees

Ingredients:

1/2 Cup soy bean flour	1/2 Tsp. salt	1 Tsp. baking powder
1/2 Cup fresh blueberries	1/3 Cup water	2 eggs

Instructions:

Mix dry ingredients. Separate eggs. Beat egg yolks with water, add to dry ingredients. Fold in stiffly beaten egg whites. Fold in blueberries. Bake at 375 for 20 minutes. 6 muffins= 315 calories; 1 muffin = 53 calories.

Blueberry Upside Down Cake

Submitted by: Ruby Parker
From: Hartly, DE
Bake Time: 30 minutes
Bake Temperature: 350 degrees

Ingredients:

1/2 Cup sugar	1 Cup flour	2 1/2 Cups blueberries
3 Tbls. butter	1 1/2 Tsp. baking powder	3 Tbls. butter
1 egg	1/2 Tsp. salt	3/4 Cup brown sugar
6 drops of vanilla	1/4 Cup milk	

Instructions:

Cream together 3 tablespoons butter, sugar and egg; stir in vanilla. Sift flour, baking powder and salt; add alternately with the milk to creamed mixture. In a buttered 8 by 8 pan, melt 3 tablespoons butter; sprinkle in the 3/4 cup brown sugar; allow to melt together. Spread blueberries over top of this mixture; then pour cake batter over the berries. Bake at 350 degrees for 30 minutes. Remove from oven and turn upside down on plate and serve with cream.

Blueberry Pie

Submitted by: J. Britton
From: Wilmington, DE
Bake Time: 10 minutes
Bake Temperature: 350 degrees

Ingredients:

1 1/4 Cups graham crackers   2 1/2 Tbls. Cornstarch   1/4 Tsp. nutmeg
2 Tbls. butter   6 Cups blueberries   1/2 Cup orange juice
1 Tbls. honey   1/2 Cup sugar

Instructions:

Preheat oven to 350 degrees. With spray, coast a 9 inch pie plate. In a bowl, mix crumbs and butter. Press into pie plate. Bake 8 to 10 minutes, until firm. Let cool. In a saucepan, bring 3 cups berries, the sugar, nutmeg, and 1/4 cup orange juice to a boil; boil 4 minutes, stirring. Stir cornstarch into remaining orange juice. Stir into berry mixture; boil 1 minute, stirring constantly with heat-resistant rubber spatula, until very thick. Spread filling in crust. Top with remaining berries; press into filling. Refrigerate 4 hours. Garnish with confectioners' sugar and orange zest.

Lemon-Blueberry Chess Squares

Submitted by: Janice Atchley
From: Wilmington, DE
Bake Time: 55 minutes
Bake Temperature: 325 degrees

Ingredients:

2 Cups blueberries   zest of one lemon, divided   1/2 Cup butter
1 Tbls. sugar   1 box yellow cake mix   4 eggs
2 3/4 Cups confectioners' sugar   (1) 8 Oz. pkg. cream cheese   2 Tbls. lemon juice

Instructions:

In a bowl, combine blueberries with sugar, lemon juice and half of the lemon zest. Set aside. Mix cake mix with 1 egg, softened butter and the remaining lemon zest. Pat into a buttered and floured 13 by 9 inch baking pan. In a mixing bowl, with a hand held mixer, beat confectioners' sugar with cream cheese and remaining 3 eggs until smooth. Gently stir in blueberry mixture; pour over cake. Bake in a preheated oven for about 45 to 55 minutes, or until browned. Let cool completely; chill and cut into squares. Serve with fresh blueberries and a dusting of confectioners' sugar, if desired. Store in the refrigerator.

Blueberry Buckle

Submitted by: Mary Lou Spicer
From: Seaford, DE
Bake Time: 60 minutes
Bake Temperature: 375 degrees

Ingredients:

2 Cups blueberries   3 Tbls. Butter   1 Tsp. baking powder
1 Tbls. lemon juice   1/2 Cup milk   1/4 Tsp. salt
1 Tbls. cornstarch   3/4 Cup sugar   1 Cup flour
 3/4 Cup sugar   1 Cup boiling water

Instructions:

Line a greased 8 by 8 pan with berries. Sprinkle with lemon juice. Cream 3/4 cup sugar and butter. In a separate bowl, sift together flour, baking powder, and salt. Add alternately with milk to butter and sugar. Spread over berries. Mix 3/4 cup sugar and cornstarch. Sprinkle over batter. Pour boiling water over top. Bake at 375 for 1 hour. Best served warm.

Blueberry Cobbler

Submitted by: Ruth Adams
From: Bridgeville, DE
Bake Time: 45 minutes
Bake Temperature: 375 degrees

Ingredients:

1/2 Cup butter   4 Tsp. baking powder   1 Quart blueberries
1 Cup sugar   1/4 Tsp. salt   1 1/2 Cups water
2 Cups cake flour   1 Cup milk

Instructions:

Cream 1/2 cup butter, 1 cup sugar, 2 cups cake flour, 4 teaspoons baking powder, and 1/4 teaspoon salt. Add 1 cup milk and beat well. Pour in pan. Put 1 quart of blueberries on top of batter. Pour 1 1/2 cups water on all. Sprinkle sugar on top and bake 45 minutes. This will fill two Pyrex cake pans or one large dish.

FLORIDA

"My family loves this recipe and I also enjoy sharing with friends. When we lived in Maine many years ago, we had a wild blueberry patch. Now I buy blueberries at our local grocery store when they are available. Earlier this fall, I bought blueberries that came from Abbotsford, British Columbia- a long way from Florida!..."

Blueberry Crumb Pudding

Submitted by: Joan Vest
From: Bell, FL
Bake Time: 30 minutes
Bake Temperature: 350 degrees

Ingredients:

1 Cup Zwieback crumbs   3 Tbls. butter   1/4 Cup sugar
2 Cups fresh blueberries   1/4 Tsp. cinnamon

Instructions:

Combine crumbs, sugar, and cinnamon; cut in butter. Place one cup blueberries in a 10 by 6 by 11/2 inch baking dish; cover with half the crumb mixture. Repeat layers. Bake at 350 for 30 minutes. Cut into squares and serve warm with ice cream. Makes 6 servings.

Blueberry Muffin Cake

Submitted by: Shirley Skinner
From: Spring Hill, FL

Ingredients:
2 pkg. blueberry muffin mix (Martha White, Jiffy or other)
(1) 3.5 Oz. pkg. cheesecake pudding mix
(1) 21 Oz. can blueberry pie filling
1 raised bottom torte pan (one that will form hollow for totte) - approximately 11" in diameter

Instructions:
Mix blueberry muffin mix as directed on package. Bake per instructions. Set aside to cool. Remove from pan and place on serving plate. Mix pudding mix using half the milk stated on package. Let set for 2 to 3 minutes and then pour into center of baked blueberry muffin cake. Top with the blueberry pie filling- about 2/3 of the can. This can be topped with a spoonful of whipped cream if desired. This cake holds well for several days in the refrigerator.

Blueberry Jam Cake

Submitted by: Janie Brackin
From: Lake Wales, FL
Bake Time: 40 minutes
Bake Temperature: 350 degrees

Ingredients:

1 Cup butter	4 eggs	1/2 Tsp. Cinnamon
1/4 Tsp. nutmeg	1 Cup sugar	1 Tsp. baking soda
1/4 Tsp. cloves	1 Cup blueberry jam	1 Cup brown sugar
3 Cups flour	1 Cup buttermilk	

Instructions:
Preheat oven to 350. Grease and flour three 9-inch cake pans. Stir baking soda into buttermilk. Set aside. In a large bowl, medium speed, cream butter and white sugar until fluffy; add brown sugar, beating until light. Add eggs, one at a time, beating. In a medium bowl combine flour, cinnamon, cloves, and nutmeg. Add buttermilk to egg mixture also with mixed flour. Then blend in blueberry jam. Pour into floured cake pans. Then bake 30 to 40 minutes or until a toothpick inserted into cake comes out clean. Cool 12 minutes then remove from pan. Create a cream cheese icing by combining a 3 ounce package of cream cheese, 6 tablespoons butter, 1 teaspoon vanilla, and one tablespoon milk in one bowl. Add in 2 or 3 cups powdered sugar to thicken texture. Mix until creamy. Spread on cake and decorate with fresh blueberries.

Blueberry Cooler

Submitted by: Mary Maurer
From: Kissimmee, FL

Ingredients:
1 Pint blueberries   2 Cups milk   1 banana
4 to 5 scoops vanilla ice cream

Instructions:
Put all ingredients in a blender and mix at high speed for about 1 minute. Serve in tall glasses immediately.

Banana Blueberry Mini Loaves

Submitted by: Muriel Jewett
From: Clearwater, FL
Bake Time: 50 minutes
Bake Temperature: 350 degrees

Ingredients:
1 Cup sugar	1 Tsp. vanilla	1/2 Tsp. salt
1/2 Cup oil	2 eggs	1 Cup fresh blueberries
1 Cup mashed banana	2 Cup flour	1/2 Cup plain yogurt
1 Tsp. baking soda		

Instructions:
Heat oven to 350. Grease and flour bottoms only of 3 or 4 mini loaf pans. In a large bowl, beat sugar and oil. Add bananas, yogurt, vanilla, and eggs. Blend well. Add flour, soda, and salt. Stir just until all moistened. Gently fold in blueberries. Pour into pans. Bake at 350 for 40 to 50 minutes or until a toothpick comes out clean. Cool for 5 minutes and then remove from pans and cool completely. The bread can be baked in a 9 by 5 loaf pan. If so, bake at 350 for 60 to 70 minutes. Canned blueberries can also be used in place of fresh ones.

Blueberry Kramm

Submitted by: Wylene Tripp
From: Zephyrhills, FL

Ingredients:

1 Pint blueberries	4 Cups water	1 Cup sugar
1 Tbls. lemon juice	4 Tbls. cornstarch	1/4 Cup cold water
Whipped cream or heavy cream		

Instructions:

Cook blueberries in water for 15 minutes. Strain. Add sugar and lemon juice to blueberry liquid. Make a past of the cornstarch and water. Add to juices. Cook stirring constantly for 3 to 4 minutes or until clear. Pour into bowl to set. Serve cold with heavy cream or whipped cream, or stir 1 cup sour cream into the mixture.

Blueberry Butterscotch Squares

Submitted by: Helene Spencer
From: Sarasota, FL
Bake Time: 30 minutes
Bake Temperature: 350 degrees

Ingredients:

3/4 Cup butter	1 1/2 Cup flour	2 Cups blueberries
1 Cup sugar	1/2 Tsp. salt	2 eggs
1/4 Tsp. baking soda	(5) 1/2 Oz. butterscotch chips	1/4 Cup chopped pecans
2 Tbls. brown sugar		

Instructions:

Cream butter, sugar, and the eggs together in a bowl. Add the flour, salt, and baking soda to the butter mixture. Fold in the blueberries. Spread mixture in a well greased 13 by 9 inch pan. Sprinkle 1/2 of the 5 1/2 ounce bag of butterscotch chips, 1/2 cup chopped nuts, and the brown sugar. Bake for 30 minutes at 350 degrees. When cool, cut into squares.

Rainbow Cake

Submitted by: Emily Sinay
From: Longwood, FL

Ingredients:

3 Oz. pkg. Jell-O- any flavor   3 Oz. pkg. blueberry Jell-O   1 container Cool Whip
2 Cups boiling water   2 baked 8 or 9 inch white cake layers, cooled

Instructions:

Place cake layers, top side up, in 2 clean layer pans; prick each cake with a fork at 1/2 inch intervals. Dissolve each flavor gelatin separately in 1 cup of the boiling water. Carefully pour the blueberry Jell-O over one cake layer and the other flavor gelatin over the second cake layer. Chill 3 to 4 hours. Dip one cake pan in warm water for 10 seconds; unmold and place on serving plate. Top with 1 cup of the whipped topping. Unmold second cake layer; place carefully on first layer. Frost top and sides with remaining whipped topping. Chill. Garnish the top with a handful of fresh blueberries.

Quick Blueberry Cream Cheese Cake

Submitted by: Jerrie Watson
From: Jacksonville, FL
Bake Time: 58 minutes
Bake Temperature: 350 degrees

Ingredients:

1 Cup blueberries   1/2 Cup sugar   3/4 Cup chopped pecans
1/2 Cup Crisco Oil (minus 2 Tsp.)   (1) 8 Oz. pkg. cream cheese   3 eggs
1 box Duncan Hines Butter Cake Mix

Instructions:

Combine the cake mix, sugar, Crisco and the cream cheese. The cream cheese should be softened. Beat the three eggs and add them into the cream cheese mixture. Beat the mixture for two minutes. Fold in blueberries and nuts. Mix well. Pour batter into a greased pan (bunt or other). Bake at 350 degrees for 58 minutes.

Blueberry Nut Bread

Submitted by: Nancy Schlegel
From: Delray Beach, FL
Bake Time: 60 minutes
Bake Temperature: 350 degrees

Ingredients:

1 Pint blueberries
1/2 Cup chopped nuts
1 Cup brown sugar
4 Tsp. baking powder (double acting)

1 Tsp. salt
3 Cups flour
1/3 Cup butter

1 Cup milk
2 eggs

Instructions:

Preheat oven to 350. Grease well and flour lightly, one 9 1/2 by 5 1/2 by 3 inch loaf pan. Stem, wash and dry fresh blueberries. Defrost to separate frozen ones. Sprinkle berries with a small amount of flour or sugar, toss lightly. Set aside. Sift and combine all dry ingredients, except sugar, in a large bowl. In a small bowl beat eggs and sugar together. Add milk and butter. Make a well in dry ingredients, and pour liquid into it. Combine with a few swift strokes (an extra large spoon with a 4 inch bowl length works well) just enough to moisten everything. Lightly fold in blueberries and nuts. Fill pan(s) about half full. Bake for 1 hour and test with skewers. Bake for 10 minutes longer if not golden brown. Cool 5-7 minutes before removing from the pan. Keeps up to six months in the freezer.

GEORGIA

"We grow lots of blueberries in Georgia. We have our own bushes and enjoy eating them right off the bush. I wanted to share this blueberry recipe with you which is an old one never less a good one, used over and over when my daughters were still at home and growing up..."

Blueberry Crunch

Submitted by: Rose Cochran
From: Moultrie, GA
Bake Time: 60 minutes
Bake Temperature: 350 degrees

Ingredients:
20 Oz. crushed pineapple with juice 1 box yellow cake mix   2 1/2 Cup blueberries
2 sticks butter   1 Cup sugar   1 Cup chopped pecans

Instructions:
Pour pineapple into 9 by 13 inch buttered dish. Add blueberries. Sprinkle 3/4 cup sugar over fruit. Sprinkle cake mix over sugar. Pour melted butter over cake mix. Sprinkle chopped nuts and 1/4 cup sugar on top. Bake at 350 for 45 minutes to 1 hour. Depending on the brand of yellow cake mix used more or less sugar can be used.

Refreshing Blueberry Dessert

Submitted by: Nana Bearden
From: Blue Ridge, GA

Ingredients:
1 Cup fresh blueberries   1/2 Cup sour cream   1/2 Cup light brown sugar

Instructions:
Wash blueberries and set aside. Mix the sour cream and brown sugar together, until smooth. Fold the blueberries into the sour cream/ sugar mixture. Place in a pretty bowl with a cover and refrigerate until flavors can blend.

Blueberry Flag Cake

Submitted by: Dwyer Morrow
From: Acworth, GA

Ingredients:
(1) 12 Oz. tub whipped topping   2 Pints strawberries   1 1/2 Cup blueberries
1 pkg. frozen pound cake (Sarah Lee's) recommended about 11 Oz.

Instructions:
Slice 1 cup of the strawberries; set aside. Halve remaining strawberries; set aside. Line bottom of a 12 by 8 inch baking dish with cake slices. Top with 1 cup sliced strawberries, one cup blueberries and all of the whipped topping. Place strawberry halves and remaining 1/3 cup of blueberries on whipped topping to create a flag design. Refrigerate until ready to serve. Makes 15 servings.

Ice Box Blueberry Pie (creates 2 pies)

Submitted by: Geri Land
From: McDonough, GA

Ingredients:

2 cooked pie shells　　　　　(1) 8Oz. pkg. cream cheese　　　1/2 Cup sugar
1 Pound confectioners' sugar　　3 Tbl. Cornstarch　　　　　　　Chopped pecans
(1) 15 Oz. can blueberries with juice
1 large box Dream Whip (both envelopes)

Instructions:

Prick pie shells with a fork. Cover bottom with nuts (optional crunch). Cream the cheese with the powdered sugar. Fold in both Dream Whips, according to directions on the box. Pour into pie shell. Top with the following blueberry mixture: Drain juice from blueberries into saucepan. Add cornstarch and the 1/2 cup sugar, stirring constantly until thick. Remove from heat and let cool "before" adding the berries. Split the topping over both pies and chill well before serving. Do not freeze.

Betty's Blueberry Cake

Submitted by: Carl Hulsey
From: Douglas, GA

Ingredients:

1 box Duncan Hines cake mix　　(1) 8 Oz. pkg. cream cheese　　1 Cup sugar
1 tub Cool Whip　　　　　　　　1 Can blueberry pie filling

Instructions:

Make the cake according to directions on the box. Make four round layers and bake. Combine sugar, cheese, and cool whip. Beat well. Ice layers, one at a time, spreading pie filling over cheese mixture. Finish top layer the same adding an edge of cream mixture around top layer, filling in with pie mixture. Keep refrigerated. This cake is so rich it is also suggested to create two layers instead of four, to be easier to divide among company.

Blueberry Cherry Crunch

Submitted by: Sharon Padgett
From: Folkston, GA
Bake Time: 30 minutes
Bake Temperature: 350 degrees

Ingredients:

1 yellow or lemon box cake mix
1/3 Cup sugar

1 Cup canned pitted cherries
3 Cups blueberries

1 Can crushed pineapple
1/2 stick butter

Instructions:

Spray 9 by 13 baking dish with non-stick cooking spray. Pour can of undrained pineapple on bottom. Scatter blueberries over the top of the pineapple. Sprinkle dry cake mix over the berries. Dot butter over the top of cake mix. Sprinkle 1/3 cup sugar over the butter. Bake at 350 for 15 minutes. Take a fork and stick in cake to allow juice to bubble up. Bake 10 to 15 minutes more or until the top is lightly browned.

Hawaii

" We don't have actual blueberries in Hawaii, but we do have a close cousin of the blueberry and cranberry called the ohelo berry. It is about the same size as a blueberry but can be red or yellow and grows on low bushes on the big island around where we live. When our son picks ohelo berries by our house, he always tosses the first one to Pele, as a tribute to the old ways in Hawaii, where people would give offerings for the goddess of the volcano. We live within 2 miles of the most active volcano on earth, Kilauea, and on the flank of another dormant volcano, Mauna Loa. It is good to respect what is around us…"

Hawaiian Blueberry Pie

Submitted by: Christine Paul
From: Kihei, HI
Bake Time: 40 minutes
Bake Temperature: 400 degrees

Ingredients:

4 Cups blueberries
1/2 Tsp. grated lemon peel
(1) 10.5 Oz. jar of pineapple topping

2 1/2 Tbls. tapioca
Pastry for 9 inch pie plate, 2 crusts

Butter

Instructions:

In a large bowl, combine pineapple topping, tapioca, and grated lemon peel; fold in blueberries. Fill pastry-lined, 9 inch pie plate and dot with butter. Adjust top crust and bake at 400 for 40 minutes. If pineapple topping comes in a 12 ounce jar that is fine.

Ohelo Berry Jam

Submitted by: · Lisa Barnard-Lapointe
From: Volcano, HI

Ingredients:
1 Cup water   4 Cups sugar
8 Cups Ohelo Berries (the Hawaiian "blueberry" as blueberries are not native to HI)

Instructions:
Pick over berries, discarding those that seem very dry, as they will not soften during cooking. Add water and cook, stirring frequently, until berries are tender (20 to 30 minutes). Add sugar, bring quickly to a boil, and cook for 5 to 10 more minutes.

Blueberry Semifreddo

Submitted by: Jewell Cote
From: Kamuela, HI

Ingredients:
4 Cups blueberries	1 Cup lowfat milk 1%	1/4 Cup honey
1/2 Cup sugar	1 Tbls. flour	Ricotta cheese
1/8 Tsp. allspice	3/4 Tsp. Vanilla	(1) 15 oz. container part-

skim cottage cheese and 1/3 Cup reduced fat sour cream

Instructions:
In a medium saucepan, combine the blueberries, sugar, allspice, and simmer over moderate heat for 5 minutes or until lightly thickened. Cool to room temperature. Meanwhile, in a small saucepan, whisk the milk into the flour. Cook, stirring, for 5 minutes or until the mixture is lightly thickened. Cool to room temperature, then transfer to a food processor. Add the ricotta, sour cream, honey, and vanilla, and process until smooth. Transfer to a bowl and fold in 2 cups of the blueberry sauce. Refrigerate the remaining sauce until serving time. Line a 9 by 5 inch glass loaf pan with plastic wrap, leaving a two inch overhang. Spoon the blueberry ricotta mixture into the pan, smoothing the top. Cover with plastic wrap and freeze for 4 hours. To serve, let stand for 30 minutes at room temperature, then unmold onto a serving platter. Cut into slices and serve with reserved blueberry sauce. Serves 8.

Creamy Blueberry Mousse With Sauce

Submitted by: Jewell Cote
From: Kamuela, HI
Bake Time: 5 minutes
Bake Temperature: medium heat

Ingredients:

2 1/2 Cup blueberries
1 pkg. unflavored gelatin
2 1/2 Tbls. sugar
(1) 8 Oz. carton low-fat vanilla yogurt

1 Tsp. Cornstarch
4 oz. Neufchatel cheese
3/4 Cup plus 2 Tbls. skim milk

1/4 Cup sugar
1/4 Cup water

Instructions:

Combine blueberries, 1/4 cup sugar, water, and cornstarch in a medium saucepan. Cook over medium heat, stirring constantly, 5 minutes or until sugar dissolves and mixture is thickened. Remove from heat, and let cool slightly. Cover blueberry mixture and chill thoroughly. Sprinkle gelatin over milk in a small saucepan; let stand one minute. Cook mixture over low heat, stirring constantly, until gelatin dissolves. Remove pan from heat; set aside, and let cool. Position knife blade in food processor bowl. Add yogurt, cheese and 2 1/2 tablespoons sugar; process 1 minute. Add gelatin mixture to yogurt mixture, and process 1 minute or until mixture is smooth. Spoon yogurt mixture into a 3 cup mold coated with vegetable spray. Cover and chill until firm. Unmold mousse on top a serving plate. Serve with blueberry sauce. Makes 6 servings.

Peek-A-Blue berry farm

IDAHO

"Greetings from Northern Idaho. I grew up in MN and picked wild blueberries every summer. Out here in Idaho we only have huckleberries, which are not as good as blueberries in my opinion…"

"Good luck in your recipe collection. This one comes from an 85 year old that wants to encourage young people that have interesting hobbies…"

Blueberry Surprise

Submitted by: Mrs. Edward Mower
From: Nampa, ID
Bake Time: 60 minutes
Bake Temperature: 350 degrees

Ingredients:

2 Cups blueberries   1 pkg. white cake mix   3/4 Cup butter
1/3 Cup sugar   1/2 Cup chopped walnuts   1 can crushed pineapple

Instructions:

Preheat oven to 350. Place the berries, sugar, undrained pineapple, cake mix and walnuts in a 9 by 13 inch pan. Melt the butter and drizzle the batter. Bake about one hour, or until a toothpick comes out clean. Serve warm or cold.

Blueberry and Cheese Cake

Submitted by: Barbara Scarsella
From: Dalton Gardens, ID
Bake Time: 30 minutes
Bake Temperature: 375 degrees

Ingredients:

1/3 Cup butter	1 1/2 Cups flour	1 Tbls. lemon juice
1/3 Cup sugar	2 Tbls. Sugar	1 Cup flour
2 eggs	1 Tbls. baking powder	1/4 Cup sugar
3/4 Cup milk	2 Cups fresh blueberries	1/4 Cup butter
(1) 3 Oz. pkg. cream cheese	Cinnamon	

Instructions:

Preheat oven to 375. Cream butter and 1/3 cup sugar until fluffy. Add eggs and beat well. Add milk and 1 1/2 cup flour mixed with baking powder. Add one cup blueberries to the batter and mix well. Pour into buttered 9 inch cake pan and sprinkle remaining blueberries on top. Combine cream cheese, sugar, and lemon juice and spread over the top of the berries. Combine remaining flour, sugar, butter, and 1/4 teaspoon cinnamon together until crumbly. Sprinkle the crumb mixture over the top of the cake. Bake for 30 minutes.

Idaho Blueberry Dessert

Submitted by: Nancy Porter
From: Kuna, ID
Bake Time: 15 minutes
Bake Temperature: 350 degrees

Ingredients:

1 Cup flour	1/2 Cup nuts	1/2 Cup butter
1/2 Cup Cool Whip	3/4 Cup confectioners' sugar	1 Cup sugar
(1) 3 Oz. pkg. cream cheese	2 Tbls. sugar	1 Tbls. Vanilla
1 Cup water	(1) 3 Oz. pkg. lemon Jell-O	1 Cup blueberries
Dash of salt		

Instructions:

Mix the flour, butter, 2 tablespoons sugar, and nuts together. Press into a 9 by 13 inch pan and bake at 350 for 15 minutes. Cool. Mix the cream cheese, vanilla, powdered sugar, and a dash of salt together. Add the Cool Whip to the cream cheese mixture and place on crust. Chill while making third layer. Heat the berries, water, and sugar. Mix and bring to a boil. Add the lemon Jell-O to berry mix. Stir to dissolve and cool. Place on top of the cream cheese mixture. Cool.

Blueberry Streusel Bread

Submitted by: Anna Doan
From: Kuna, ID
Bake Time: 45 minutes
Bake Temperature: 375 degrees

Ingredients:

4 Tbls. butter
3/4 Cup sugar
1 egg
1/2 Cup milk

2 Cups flour
2 Tsp. baking powder
1 Tsp. salt
2 Cups fresh blueberries

1/2 Cup sugar
1/3 Cup flour
4 Tbls. butter
1 Tsp. cinnamon

Instructions:

Preheat oven to 375. Grease a 9-inch square-baking pan. With electric mixer cream 3/4 cup sugar and 4 tablespoons butter until light and fluffy. Add egg. Blend until completely combined, and then blend in milk. Sift 2 cups flour, baking powder, and salt and into creamed mixture just enough to blend all ingredients. Add blueberries and stir. Pour into baking dish. To create topping, place flour, sugar, cinnamon, and butter into a bowl. Blend together by cutting with knives or pastry blender until mix resembles coarse crumbs. Sprinkle over batter in pan. Bake about 45 minutes. Can be eaten warm or cold or with ice cream.

Peek-A-Blue berry farm

*I*LLINOIS

"'Twas a delight to see your note asking for blueberry recipes! You are the same age as the girls in Home Ec classes I taught way back in the mid 1950's..."

" The enclosed recipe appeared in a magazine many years ago and was titled Betty Ford's Blue-Bana Bread. A brief article stated that the family of former President Jerry Ford enjoyed then First Lady Betty Ford's Blue-Bana Bread when residents of the White House."

Blueberry Crisp

Submitted by:	Dianne Meyers
From:	Lakemoor, IL
Bake Time:	45 minutes
Bake Temperature:	375 degrees

Ingredients:

4 Cups blueberries	2 Tbls. lemon juice	1/4 Cup brown sugar
1/2 Cup sugar	1/2 Cup rolled oats	2 Tbls. cornstarch
2 Tbls. chopped toasted walnuts	1/2 Cup flour	6 Tsp. butter

Instructions:

Preheat the oven to 375. Coat a 1-quart casserole with nonstick spray. In a large bowl, mix the blueberries, sugar, cornstarch, and lemon juice. Spoon into the prepared casserole. In the same bowl, mix the oats, flour, brown sugar, and walnuts. With a fork or pastry blender, beat in butter until the mixture resembles a course meal. Sprinkle over the berry mixture. Bake for 45 minutes, or until lightly browned and bubbling. Creates 8 servings.

Blueberry Salad

Submitted by: Mrs. Zehner
From: Western Springs, IL

Ingredients:

2 heads Bibb or Leaf Lettuce
1 Cup fresh blueberries
1 1/2 Oz. toasted sunflower seeds
2 Cups Vegetable Oil
2 - 3 Tablespoons poppy seeds

1 pkg. toasted pecans
1 1/2 Tsp.Salt
1 to 2-sliced bananas
1/2 Cup Monterey Jack cheese, shredded
Dresssing: one small white onion minced

2/3 Cup White Vinegar
1 1/2 Cups Sugar
2 Tsp. Dry Mustard

Instructions:

Tear lettuce into bite size pieces. Toss with remaining ingredients. Then mix the following to create a dressing: one small white onion (minced), 2/3 cup white vinegar, 1 1/2 teaspoon salt, 2 to 3 tablespoons poppy seeds, 1 1/2 cup sugar, 2 teaspoons dry mustard, and two cups vegetable oil.

Blueberry Glaze Pie

Submitted by: Mary Schellenger
From: Zion, IL
Bake Time: 15 minutes
Bake Temperature: 425 degrees

Ingredients:

1 1/2 Cup flour
1 1/2 Tsp. sugar
Dash of nutmeg
3 Tbls. lemon juice

3 Tbls. cornstarch
2 Tbls. milk
1 Tsp. salt
3/4 Cup sugar

3/4 Cup water
1 Pint blueberries
1/2 Cup vegetable oil

Instructions:

Mix 1 1/2 cup flour, 1 1/2 teaspoon sugar, and salt in a piepan. Make a well in the middle. Mix vegetable oil and milk together with fork. Add to flour mix. Mix with fork and fingers until it holds together fairly well. Press to line bottom and sides of pan. Bake at 425 12 to 15 minutes. In a saucepan, mix sugar and cornstarch. Add one cup of the berries, the water, lemon juice, and nutmeg. Blend and cook, stirring, until thickened. Fold in the rest of the berries and pour into the pie shell. Chill to set. Top with whipped cream or Cool Whip.

Betty Ford's Blue-Bana Bread

Submitted by: Ellen Kucera
From: Lemont, IL
Bake Time: 50 minutes
Bake Temperature: 325 degrees

Ingredients:

2 Cups blueberries
4 Cups flour
1 Cup butter
2 Cups sugar

4 eggs
2 Tsp. vanilla
1 Tbls. ground allspice
5 medium ripe bananas, mashed

2 Tsp. baking soda
1 Tsp. baking powder
1/2 Tsp. salt

Instructions:

Preheat oven to 325. Grease and flour (2) 9 x 5 inch loaf pans. Toss blueberries with two tablespoons flour. Set aside. In a large bowl, with mixer at medium speed, cream butter and sugar. Add eggs, bananas and vanilla. Beat at low speed until blended. Add remaining flour, allspice, soda, powder, and salt. Beat at low speed until well blended. Fold in blueberries. Pour into pans. Bake 45 to 50 minutes. Cool in pans on wire rack for 10 minutes. Remove from pans and cool for at least 20 minutes to serve warm; or cool completely to serve later.

Blueberry- Apple Dessert

Submitted by: Betty Peters
From: Gibson City, IL
Bake Time: 40 minutes
Bake Temperature: 350 degrees

Ingredients:

1 pint blueberries
2 Cups peeled tart apples, sliced
1 Tbls. lemon juice
1/2 Cup brown sugar

3/4 Cup sugar
1 Tsp. baking powder
3/4 Tsp. salt
1 Cup flour

1 egg, slightly beaten
1/3 Cup melted butter
1/2 Tsp. cinnamon

Instructions:

Grease the baking dish. Put in apples and blueberries. Sprinkle with lemon juice and brown sugar. Mix together the sugars, flour, baking powder, and salt. Stir in egg. Mix until crumbly. Sprinkle over fruit. Top with butter and cinnamon. Bake at 350 for 35 to 40 minutes.

Blueberry Ice Cream Pie

Submitted by: Lulu Zimmerman
From: Eureka, IL
Bake Time: 375/475 degrees
Bake Temperature: 10 minutes

Ingredients:

1 Quart vanilla ice cream   1/4 Tsp. cream of tartar   1/4 Cup butter, softened
3 egg whites   1/2 Tsp. vanilla   1/4 Cup sugar
6 Tbls. sugar   2 Tsp. Cornstarch   1/2 Cup Water
1 Tsp. Lemon Juice and Rind   1 pint of Blueberries   Dash of salt
1 1/3 Cups fine graham cracker crumbs
1 1/2 Cup blueberry sauce* = 1/2 Cup Sugar

Instructions:

Mix crumbs, butter, and sugar until crumbly. Press firmly in a unbuttered 9 inch pie plate. Bake in a moderate oven- about 375 degrees for 8 minutes. Cool. Spread half of ice cream in cooled crumb shell. Mix remaining ice cream with blueberry sauce. Fill pie. Freeze until firm. Allow egg whites to reach room temperature. Beat egg whites, vanilla, and cream of tartar until soft peaks form. Gradually add 6 tablespoons sugar, beating until stiff and glossy. Cover entire surface of ice cream with meringue, sealing to edges. Place pie plate on board and bake at 475 for 2 to 3 minutes, or until lightly browned. Serve at once. Makes 6 to 8 servings.

*Blueberry Sauce

Combine 1/2 cup sugar, 2 teaspoons cornstarch, and a dash of salt. Stir in 1/2 cup water. Add a pint of blueberries. Bring to a boil; simmer until clear and thickened, about 4 minutes. Remove from heat and add lemon juice and rind. Chill.

INDIANA

"I think it's wonderful that you are collecting blueberry recipes. If you have a computer, go to Wal-Mart and look in the small appliance department and ask for 1 Million Recipes. It's the largest cookbook on CD- there are over 6000 blueberry recipes. Happy collecting!"

"My mother gave me this recipe in 1962. Everyone loves it, I always have it for all holidays. In our town, we have a blueberry parade and festival every Labor Day. It's 3 days every year…"

Tangy Crock-Pot Blueberry Sauce

Submitted by: Joan Miller
From: Lowell, IN

Ingredients:

8 Cups partially thawed blueberries   1 Tsp. cinnamon   2 Cups sugar
1/4 Tsp. allspice   2 Tbls. lemon juice   1/4 Tsp. cloves

Instructions:

Combine in a crock-pot. Stir occasionally. Cover and cook on high 1-2 hours. Turn on low and cook 6-8 hours. Sauce should be thick when done. Can be used on biscuits, cake, waffles, pancakes, and more.

Blueberry Brancakes

Submitted by: Arbutus Eversoll
From: Petersburg, IN

Ingredients:

3 Cups flour, sifted   1 Tsp. baking soda   1/4 Cup oil
1 Cup all bran cereal   2 1/2 Cup milk   Grated rind of one orange
4 Tsp. baking powder   1/3 Cup honey   2 Cups blueberries
2 eggs

Instructions:

Mix flour, bran, baking powder, and baking soda. Mix in the two well beaten eggs, milk, honey, and oil. Add liquid all at once to dry ingredients. Add orange rind and blueberries and stir only until all dry ingredients are moistened. Pour about 1/3 cup of the batter for each pancake on top a preheated slightly- greased skillet. Bake on one side until bubbles appear. Turn and brown on other side.

Blueberry Drop Cookies

Submitted by: Marlene Kurtz
From: Fort Wayne, IN
Bake Time: 12 minutes
Bake Temperature: 375 degrees

Ingredients:

1/2 Cup sugar   1/4 Cup sour milk   1 Cup fresh blueberries
1/2 Cup light brown sugar   2 1/4 Cups flour   1 Tsp. baking powder
1/2 Cup butter   2 Tsp. honey   1 Tsp. vanilla
1 egg   1/2 Tsp. baking soda

Instructions:

Heat oven to 375. Cream together the sugars, butter, honey, egg, and sour milk. Sift flour, baking powder, and baking soda. Mix together. Fold in vanilla and blueberries. Drop by spoonfuls onto a greased cookie sheet. Bake 10 to 12 minutes at 375.

Raspberry and Blueberry Topping

Submitted by: Anna Mowan
From: Spencerville, IN

Ingredients:

1 2/3 Cup water   2 Tbls. Cornstarch   2/3 Cup water
Red food coloring   2 Tbls. white Karo Syrup   (1) 3 Oz. box raspberry
Jell-O   1/2 bag frozen store bought blueberries
1/2 bag frozen store bought raspberries

Instructions:

Mix and stir all ingredients together and cook until thickened. Remove from stove and add the raspberry Jell-O. Let cool and add the frozen berries so it will thicken. Warm in the microwave and serve over pancakes or ice cream.

Banana Blueberry Pie

Submitted by: Susan Lambright
From: Middlebury, IN

Ingredients:

(1) 8 Oz. pkg. cream cheese   2 baked 9 inch pastry shells   3/4 Cup sugar
4 medium firm bananas, sliced   2 Cups whipped topping
(1) 21 Oz. can blueberry pie filling

Instructions:

In a mixing bowl, beat cream cheese and sugar until smooth. Fold in whipped topping and bananas. Pour into two baked pastry shells. Spread with pie filling. Refrigerate for at least 30 minutes. Just before serving garnish with fresh banana slices, blueberries, and mint.

Blueberry Peach Trifle

Submitted by: Diana Hogue
From: Akron, In

Ingredients:

1 1/2 Cups cold water
(2) 9 Oz. cans sliced peaches
1 can sweetened condensed milk

1 Pint whipped whipping cream
4 Cups cubed pound cake
(1) 3 1/2 Oz. box vanilla instant pudding

2 Tsp. grated lemon rind
4 Cups blueberries

Instructions:

In a large bowl, combine milk, water, and lemon rind. Add pudding mix and beat until blended. Chill for 5 minutes. Fold in whipped cream. Spoon 2 cups pudding mix into 4-quart glass bowl. Top with half of cake cubes, the drained peaches, half of the pudding mixture, the rest of the cake cubes, then berries and pudding on top. Chill for 4 hours before serving.

Iced Blueberry Dessert

Submitted by: Diana Hogue
From: Akron, In

Ingredients:

1 1/2 Cups blueberries
2 Tbls. fresh orange juice
1/2 Cup soft macaroon crumbs

2 large peaches, peeled and sliced
1 Pint vanilla ice cream
Dash of cinnamon

Instructions:

Combine blueberries, orange juice, and cinnamon in a blender until smooth. Spoon 2 tablespoons of macaroon crumbs into four dessert dishes. Cover with peach slices. Top with a scoop of vanilla ice cream. Spoon blueberry topping over ice cream

Blueberry Peach Pound Cake

Submitted by: Lorraine Pflederer
From: Goshen, IN
Bake Time: 70 minutes
Bake Temperature: 350 degrees

Ingredients:

1/2 Cup butter   1/4 Cup milk   1/4 Tsp. salt
1 1/4 Cups sugar   2 1/2 Cups cake flour   3 eggs
2 Tsp. baking powder   2 Cups blueberries
2 - 1/4 Cups chopped peeled fresh peaches

Instructions:

In a mixing bowl, cream butter and sugar. Beat in eggs, one at a time. Beat in milk. Combine the flour, baking powder and salt; add to creamed mixture. Stir in peaches and blueberries. Pour in a greased and floured 10 inch fluted tube pan. Bake at 350 for 60 to 70 minutes. Cool in pan for 15 minutes. Top with confectioners' sugar. Creates 10 -12 servings.

Blueberry Chutney

Submitted by: Darlene Potts
From: Grandview, IN

Ingredients:

3 Cups blueberries   1/4 Cup chopped onion   1/3 Cup cider vinegar
(1) 3 inch cinnamon stick   1 Tbls. fresh grated ginger root 1 Tbls. Cornstarch
1/4 Tsp. salt   1/2 Cup firmly packed brown sugar

Instructions:

In a large saucepan, combine all ingredients. Bring mixture to boil over medium heat, stirring frequently. Boil one minute. Remove cinnamon stick, cool. Cover, refrigerate. Serve as a condiment with meats and cheeses.

Summer Dessert Pizza

Submitted by: Michael Bilby
From: Greenfield, IN
Bake Time: 14 minutes
Bake Temperature: 350 degrees

Ingredients:

1/4 Cup butter
1/2 Cup sugar
1 egg
(1) 4 Oz pkg. cream cheese
(1) 8 Oz. can mandarin oranges
1/3 Cup fresh blueberries
1/4 Cup confectioners' sugar

1 1/4 Cups flour
1/4 Tsp. lemon extract
1/4 Tsp. baking soda
1/4 Tsp. vanilla
1 Cup fresh sliced strawberries
1 Cup whipped topping
2 kiwifruit peeled and sliced thin

1/4 Tsp. baking powder
1/4 Tsp. salt
2 Tsp. cornstarch
1/4 Cup sugar
1/4 Cup orange juice
1 firm banana

Instructions:

In a mixing bowl, cream butter and sugar, beat in eggs and extracts. Combine flour, baking powder, baking soda, and salt, add to creamed mixture. Beat well. Cover and chill for 30 minutes. Press dough into a greased 12 inch or 14 inch pizza pan. Bake at 350 degrees for 12 to 14 minutes or till light golden brown. Cool completely, in a mixing bowl beat cream cheese and confectioners sugar until smooth. Spread over crust. Arrange fruit on top. Combine the remaining sugar, orange juice and cornstarch together in a sauce pan. Bring to a boil, stirring constantly. Boil for two minutes or until thickened. Cool to room temperature, about 30 minutes. Brush over fruit. Store in refrigerator.

IOWA

Anthocyanin is the blue pigment found in the skins of blueberries. Recent research has shown that anthocyanins are thought to help control diabetes, improve blood circulation, reduce eyestrain, and slow the effects of aging as well as improve memory skills.

Blueberry Pudding

Submitted by: Norma Kidwell
From: Council Bluffs, IA

Ingredients:

2 Cups low fat cottage cheese   granola to line a pie plate   1 Cup low fat yogurt
2 Cups blueberries   3 Tbls. lemon juice   2 Tbls. honey

Instructions:

Blend cottage cheese in blender; add yogurt, honey, and lemon juice. Fold in blueberries. Pour into a pie plate lined with granola. Refrigerate several hours before serving. Serves 8.

Automated Freedom Blueberry Bread

Submitted by: Jo Rorabaugh
From: Urbandale, IA

Ingredients:

1 pkg. yeast	2 Tbls. sugar	3 Cups bread flour
1 Tbls. butter	1/4 Cup water	1/2 Tsp. salt
(1) 16Oz. blueberries	1/4 Cup retained juice from blueberries	

Instructions:

Add the yeast, flour, salt, and sugar into pan. Put well drained berries into a 2 cup measuring cup, add 1/4 cup juice and enough water to equal 1 1/3 cups. Place into bread machine. Select white bread and push start.

Blue Witch's Brew

Submitted by: Erica Swanger
From: Tama, IA

Ingredients:

2 1/2 Cups blueberries	1/4 Cup milk	1 1/4 Cup apple juice
3/4 Tsp. ground cinnamon	1 Cup vanilla ice cream	

Instructions:

In a blender, combine blueberries, apple juice, ice cream, milk, and cinnamon until smooth. Serve immediately. Creates about 4 cups.

Blueberry Tortilla Pizza

Submitted by: Erica Swanger
From: Tama, IA

Ingredients:

1/2 Cup sliced strawberries	1 Tbls. confectioners' sugar	1 Pint fresh blueberries
1 large 10 inch flour tortilla	2 Tsp. cinnamon sugar	1 Tbls. butter
1/4 Cup toasted shredded coconut	1/2 Cup ricotta or whipped cream cheese	

Instructions:

Preheat broiler. In a small bowl, combine ricotta cheese and confectioners' sugar; set aside. In another small bowl, combine blueberries and strawberries. Arrange tortilla on a broiler pan; brush with butter and sprinkle with cinnamon sugar. Broil about 6 inches from heat source, until lightly browned, about 3 minutes. Cool slightly. Spread ricotta mixture on the tortilla; top with blueberry mixture and then sprinkle with cocoanut. Creates 4 servings. To toast coconut, place in a skillet over moderate heat until pale gold, stirring constantly.

Blueberry-Peach Cobbler

Submitted by: Edna Korn
From: Missouri, VLY, IA
Bake Time: 45 minutes
Bake Temperature: 350 degrees

Ingredients:

1 pkg. wild blueberry muffin mix	1/2 stick butter	1 Tsp. cinnamon
1/4 Cup sugar	1/2 Cup chopped pecans	1/2 Tsp. cinnamon
1 Tsp. almond extract	1/4 Cup sugar	2 cans peach pie filling

Instructions:

Preheat oven to 350. Wash blueberries; set aside to drain. In a medium size bowl, combine dry muffin mix, 1/4 cup sugar, and 1/2 teaspoon cinnamon. Cut in butter, then stir in nuts. In a 9 by 13 inch pan combine pie filling, 1/4 cup sugar, 1 teaspoon cinnamon, the almond extract, and the blueberries. Spoon crumbled muffin mix on the peach mixture. Bake at 350 for 35 to 45 minutes or until topping is golden brown. Serve with ice cream.

Julie's Blueberry Cheese Squares

Submitted by: Marlys Breese
From: Iowa City, IA

Ingredients:

1 1/2 pkg. graham crackers
2 Tbls. confectioners' sugar

1 can blueberry pie filling
(1) 8 Oz. pkg. cream cheese

1/2 Cup butter
1/4 Cup milk

Instructions:

Roll graham crackers to fine crumbs. Combine with melted butter and mix well. Reserve 1/2 cup of the crumb mixture for topping. Press the remaining crumbs into an 8 by 10 inch pan. Chill. Blend cream cheese, powdered sugar, and milk until smooth. Spread over chilled crumb layer. Spoon blueberry pie filling over cream cheese. Top with cool whip or whipped cream. Sprinkle with reserved crumbs. Chill until ready to serve.

KANSAS

"We pick blueberries at a place called Blueberry Hill, just south of our home in Overland Park, Kansas. They also have a small gift shop attached where we purchase jams and jellies and they give out wonderful blueberry recipes. You will be very busy if you try all the recipes you receive. Have fun with your project."

Blueberry Bars

Submitted by: Marilyn Fight
From: Leavenworth, KS
Bake Time: 45 minutes
Bake Temperature: 350 degrees

Ingredients:

2 sticks butter
1 3/4 Cup sugar
1 Tsp. vanilla

4 eggs
1 Cup whole wheat pastry flour
2 Cups unbleached flour

1 1/2 Tsp. baking powder
2 Cups blueberries

Instructions:

Cream together butter and sugar. Add eggs and vanilla. Add flour and baking powder. Spread in 11 by 13 inch sheet cake or jellyroll pan sprayed with oil. "Cut" dough into squares and put 4 or 5 blueberries in the center of each square. Dough will bake up around the blueberries. Bake at 350 degrees for 40 to 45 minutes. Sprinkle with powdered sugar. Cut into bars. These freeze well.

Goodbye Turkey Blueberry Salad

Submitted by: Marilyn Fight
From: Leavenworth, KS

Ingredients:

2 Cups chopped and cooked turkey   1 Cup frozen blueberries   Lettuce leaves
1 Cup diced celery   1/2 Cup chopped pecans   Paprika
1 Small sliced banana   1/2 Cup mayo   1/4 Cup plain yogurt
(1) 8 Oz. can unsweetened pineapple chunks, drained

Instructions:

Combine the turkey, celery, pineapple, banana, blueberries, and pecans. Toss lightly. Mix mayonnaise and yogurt very well. Add to turkey mixture and toss carefully. Serve on lettuce leaves and sprinkle with paprika and a few fresh blueberries.

Breakfast Surprise

Submitted by: Marilyn Fight
From: Leavenworth, KS
Bake Time: 25 minutes
Bake Temperature: 375 degrees

Ingredients:

(1) 8 Oz. box corn bread mix   1 Cup blueberries
(1) 15 Oz. can sliced peaches, drained
(1) 7 Oz. pkg. brown and serve sausage or turkey sausage

Instructions:

Prepare corn bread batter according to the package directions; set aside. Brown sausage, cut links into fourths. Arrange peaches, blueberries, and sausage in bottom of a greased 9 inch square dish. Pour corn bread batter over top. Bake at 375 degrees for 25 minutes. Cut into squares and serve upside down, with peaches, blueberries and sausage on top.

Blueberry Streusel

Submitted by: Ellen Cooper
From: Leawood, KS
Bake Time: 40 minutes
Bake Temperature: 375 degrees

Ingredients:

2 Cups flour
1/4 Tsp. salt
2 Tbls. Flour
2/3 Cup salad oil

1 Pint blueberries
2 Tbls. sugar
2 Tbls. butter
1/2 Tsp. cinnamon

3/4 Cup sugar
3 Tbls. milk
1 1/4 Tsp. salt

Instructions:

Combine the 2 cups flour, 2 tablespoons sugar, 1 1/4 teaspoon salt, salad oil and milk. Place in Pyrex 8 1/2 by 11 baking dish. Use fork to press crust into a nice even layer. Gently add blueberries to cover crust. With fork mix the remaining sugar, flour, cinnamon, and butter. Crumble on top of blueberries. Bake at 375 for 40 minutes. Can be served warm with ice cream.

Blueberry Pie

Submitted by: Dorothy Middlebusher
From: Wichita, KS
Bake Time: 45 minutes
Bake Temperature: 425 degrees

Ingredients:

1/4 Cup quick cooking tapioca
1/4 Cup granulated sugar
2 Tbls. brown sugar

2 Cups blueberries
1/4 Tsp. cinnamon
1/4 Tsp. salt

Pastry for (2) 8 inch pies
1 Tbls. lemon juice
1/2 Cup blueberry juice

Instructions:

Combine tapioca, sugars, salt, cinnamon, blueberries, blueberry juice, and lemon juice. Set aside. Pour filling into piecrust; dot with butter. Roll top crust over filling. Open slits to let steam escape. Bake at 425 for about 45 minutes.

Spiced Blueberry Jam

Submitted by: Linda Kibbe
From: Clearwater, KS

Ingredients:

3 Cups blueberries   1 Tsp. cinnamon   1 pouch liquid pectin
1 Tbls. lemon juice   1/4 Tsp. allspice   Melted Paraffin
3 1/2 Cups sugar   1/4 Tsp. ground cloves

Instructions:

Remove any stems from berries. Crush fruit 1 layer at a time. Measure 2 1/4 cups, packed solidly. If necessary, add water to make up for that amount. Pour into very large saucepan or Dutch oven. Add lemon juice, sugar, and spices; mix well. Bring to a full rolling boil and boil, stirring, one minute. Remove from heat and immediately stir in pectin. Ladle into hot jelly glasses or jars. Pour 1/8 inch hot paraffin over top. Paraffin should cling to sides of jars and contain no air bubbles. Cover with lids and store in a cool dry place.

The Blues Salad

Submitted by: Jeanette Lewis
From: Latham, KS

Ingredients:

2 Tbls. walnut oil or olive oil   2 Tbls. honey   1/4 Tsp. salt
1/4 Tsp. coarsely ground pepper   2 Cups rice vinegar   3 Cups blueberries
1 Cup blueberries   4 Cups torn mixed greens
1/2 Cup walnut halves, lightly toasted
2 Tbls. snipped chives or sliced green onion tops
1/4 Cup crumbled Stilton or blue cheese

Instructions:

To create a blueberry vinegar for the salad, place 1 1/2 cups of blueberries with 2 cups of rice vinegar in a stainless steel or enamel saucepan. Bring to a boil; reduce heat. Simmer uncovered for 3 minutes. Stir in 2 tablespoons of honey. Remove from heat. Pour mixture through a mesh strainer and let it drain into a bowl. Discard berries. Transfer strained vinegar to clean 1 quart jar or bottle. Add another1 1/2 Cup cups blueberries to the jar or bottle. Cover tightly with a nonmetallic lid. Store vinegar in a cool, dark place for up to 6 months. Before using vinegar, discard berries. To make a fresh blueberry salad, combine 1/4 cup blueberry vinegar, walnut oil, salt, and pepper in a screw-top jar. Cover; shake well. In a large bowl, combine greens, 1 cup blueberries, walnut halves, crumbled cheese, and snipped chives. Pour desired amount of dressing over salad. Toss lightly to coat. Divide salad mixture evenly among four individual bowls.

KENTUCKY

"My husband and I have blueberry bushes. We have had a lot of berries for several years. One of our chickens even ate them out of my hand. Keep cooking goodies…"

"I used this recipe quite often when our 8 children were still at home on the farm and when our blueberries were in season. There are now 36 grandchildren, my 93-year-old mom, nieces and nephews, which makes a large gang, 66 in all. It takes a lot of food too.

Blueberry Turnovers

Submitted by:	Faye Irvin
From:	Boston, KY
Bake Time:	30 minutes
Bake Temperature:	400 degrees

Ingredients:

pastry for 2 crust pie

1 Tsp. lemon juice

1/3 Cup sugar

1/8 Tsp. salt

1 Cup confectioners' sugar

2 Tbls. flour

1 Cup blueberries

1 Tbls. milk

1/2 Tsp. vanilla

Instructions:

Roll pastry to 1/8 inch thickness; cut in eight 5 inch squares. Combine blueberries, sugar, flour, lemon juice, and salt; place a heaping tablespoon of mixture on each square. Fold pastry in half diagonally; seal edges with fork. Cut slits for escape of steam. Place turnovers on baking sheet; bake at 400 degrees for 20 to 30 minutes. Frost with a mixture of the milk, vanilla, and confectioners' sugar.

Blueberry Refrigerator Jam

Submitted by: Alice Carwile
From: Leitchfield, KY

Ingredients:

1 3 Oz. pkg. lemon gelatin   4 Cups blueberries   2 Cups sugar

Instructions:

In a large saucepan, combine all three ingredients. Bring to a boil. Cook and stir for two minutes. Pour into jars and refrigerate.

Light Blueberry Sauce

Submitted by: Alice Carwile
From: Leitchfield, KY

Ingredients:

3 Tbls. orange juice   1 Cup blueberries   1 1/2 Tsp. cornstarch
2 Tbls. sugar

Instructions:

Combine all ingredients in a small saucepan. Heat over medium heat, stirring until thick. Serve over pancakes, ice cream, and cake.

Blueberry Torte

Submitted by: Della Lacy
From: Campton, KY
Bake Time: Until light brown
Bake Temperature: 250 degrees

Ingredients:

3 egg whites
1 Cup white cracker crumbs
1 Tsp. vanilla

1 can blueberry pie filling
1 Cup sugar
(1) 8 Oz. pkg. cream cheese

1 Cup Cool Whip
1/2 Cup chopped nuts
3/8 Tsp. cream of tarter

Instructions:

Beat egg whites stiff. Add vanilla, sugar, and cream of tarter. Mix well. Fold in nuts and cracker crumbs. Pour mixture into a greased 15 by 9 inch pan. Bake 250 degrees until light brown. Cool. Top with a mixture of the cream cheese and Cool Whip. Spread pie filling. Decorate top with more Cool Whip as desired. Keeps several days in the refrigerator.

Blueberry Surprise

Submitted by: Rosemary Fahey
From: Burlington, KY
Bake Time: 15 minutes
Bake Temperature: 350 degrees

Ingredients:

1 can blueberry pie filling
1 carton whipped topping
(1) 8 Oz. pkg. cream cheese

1 Cup sugar
1 stick butter
1 Cup flour

1 Tsp. vanilla
1 Cup chopped pecans
1/4 Cup brown sugar

Instructions:

Melt the butter in a oblong baking dish. Mix the flour, brown sugar, and pecans into the melted butter. Bake for 15 minutes at 350 degrees. Cool well. Mix sugar, cream cheese, whipped topping, and vanilla and pour into cooled crust. Add one can blueberry pie filling on top of the above mixture. If desired, top with a small carton if whipped topping.

Cream Cheese Fruit Bread

Submitted by: Charlene Strunk
From: Strunk, KY
Bake Time: 60 minutes
Bake Temperature: 350 degrees

Ingredients:

1 8 Oz. pkg. cream cheese	4 eggs	2 Cups fresh blueberries
1 Cup butter	1 1/2 Tsp. baking powder	1 1/2 Cup sugar
2 1/2 Cups flour	1 1/2 Tsp. vanilla	1/2 Cup coconut

Instructions:

Blend cream cheese, butter, sugar, and vanilla. Add eggs one at a time and beat well after each one is added. Gradually add flour and baking powder. Fold in blueberries and coconut. May bake in 2 loaf pans of muffin pans (greased) at 350 for 1 hour for loaf pans, or 30 to 40 minutes for muffins. Cool in pans.

Blueberry Strudel

Submitted by: Roberta Green
From: South Shore, KY
Bake Time: 40 minutes
Bake Temperature: 350 degrees

Ingredients:

2 sticks butter	2 Cups flour	1 can vanilla frosting
1 Cup sour cream	1 beaten egg	1 can blueberry Pie filling

Instructions:

Beat the butter and sour cream until well blended. Add the flour. Beat together, cover and put in the refrigerator overnight. Roll out on a floured board. Put on cookie sheet. Add thin layer of blueberry pie filling. Fold together. Pinch top and ends together. Brush with a beaten egg. Bake at 350 for 30 to 40 minutes. Glaze with frosting while still warm.

LOUISIANA

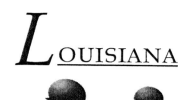

"This is from the Louisiana blueberry cook-off held yearly in Mansfield, LA which is a rural town 23 miles south of Shreveport. There are many blueberry farms in this part of Louisiana. Most are sold commercially and most are the big beautiful rabbiteye as we call them…"

Upside Down Blueberry Pudding

Submitted by: Mary Vercher
From: Youngsville, LA
Bake Time: 40 minutes
Bake Temperature: 350 degrees

Ingredients:

1 Oz. butter
1 box yellow or white cake mix
2/3 Cup water

1 egg
1 Tsp. mixed spice

2 Cups blueberries
1/2 Cup brown sugar

Instructions:

Melt butter in baking dish or cake pan. Mix in brown sugar and spread evenly. Arrange fruit on top. Mix cake mix, spice, water, and egg until smooth. Pour over fruit and bake about 40 minutes at 350 degrees. Turn out and serve with ice cream.

Blueberry Flip

Submitted by: Paula Willey
From: Pine Grove, LA

Ingredients:

1 Cup blueberries
Vanilla ice cream
Ginger ale
2 bananas, sliced
1 Cup sliced peaches

Instructions:

In a medium bowl, combine all the fruit. Spoon 1/2 of the fruit into 4 tall glasses. Top with a scoop of vanilla ice cream. Add the rest of the fruit and another scoop of ice cream. Fill the glasses with ginger ale and serve. Makes 4 servings.

Blueberry Glazed Cornish Game Hens

Submitted by: Mary
From: Stonewall, LA
Bake Time: 90 minutes
Bake Temperature: 375 degrees

Ingredients:

1 Cup cooked brown rice
2 ribs celery chopped
1/8 Tsp. poultry seasoning
1 small chopped onion
2 Tbls. butter
1/2 Tsp. grated orange rind
Melted butter
1/2 Cup blueberry syrup
Cornish Game Hens
1 1/2 Tsp. lemon juice

Instructions:

Prepare rice as directed on package, set aside. Sauté onion and celery in 1-tablespoon butter until tender. Combine vegetables, rice and poultry seasoning. Remove giblets from hens, rinse with cold water and pat dry. Sprinkle cavity with water. Stuff hens lightly with rice mixture. Place hens breast side up in baking dish. Brush with melted butter. Bake at 375 for 1 hour and 30 minutes, basting occasionally with butter. Create a blueberry glaze after putting hens in oven by combining the blueberry syrup and orange rind in saucepan and bringing to a boil. Remove from heat; add butter and lemon juice. Stir until butter melts. After the hens bake for the 90 minutes, spoon blueberry glaze over hens and bake an additional 5 minutes. Serve over additional rice if desired.

Blueberry Banana Pie

Submitted by: Linda Smith
From: Shreveport, LA

Ingredients:

1 can Eagle brand mix   fresh bananas   pint of blueberries
1/3 Cup lemon juice   3 Tbls. Confectioners' sugar (1) 9" baked crust
1 carton whipping cream   1/4 Tsp. almond extract

Instructions:

Mix the Eagle brand mix and the lemon juice together. Add whipping cream, confectioners' sugar, and almond extract to the lemon mixture. Put a small amount of mixture into a baked crust. Place bananas over the mixture and add remaining mixture. Spread blueberries over the top.

Peek-A-Blue berry farm

M AINE

"I have 5 children, 4 live out of state and I have even mailed them some for they so much enjoy fresh Maine blueberries. We have the small wild ones that you pick down on the ground. About 9 miles from here is the Bath Maine Iron Works, a big ship yard that builds many ships that have taken part in all the wars…"

"Over 98% of the wild blueberries produced in the United States are grown in Maine. Hand-held rakes that comb the berries from the bush harvest them. Blueberry fields are harvested every other year. Wild blueberries have a slightly tart, piquant flavor unlike any other berry or fruit…"

Wild Blueberry Trifle

Submitted by: Rose Moulton
From: Norway, ME

Ingredients:
4 Cups cubed pound cake
2 Tsp. grated lemon rind
(1) 3 1/2 Oz. pkg. instant vanilla pudding and pie filling mix
(1) 14 Oz. can sweetened condensed milk

1 1/2 Cups cold water
4 Cups blueberries

2 Cups whipping cream

Instructions:
In a large bowl, combine sweetened condensed milk, water, and lemon rind; mix well. Add pudding mix; beat until well blended. Chill 5 minutes. Fold in whipped cream. Spoon 2 cups pudding mixture into 4 quart glass serving bowl. Top with 1 cup cake cubes, 2 cups blueberries, then with half the remaining pudding mixture. Continue layering the remaining cake cubes then the blueberries and remaining pudding mixture. Chill at least 4 hours. Garnish as desired. Creates about 10 servings.

Blueberry-Orange Marmalade

Submitted by: Stella Brown
From: Gardiner, ME

Ingredients:

1 orange   3/4 Cup water   2 1/2 Cups sugar
1/8 Tsp. soda   (1) 6 Oz. bottle liquid pectin   4 Cups blueberries
1 lemon

Instructions:

Cut orange and lemon rinds into small pieces. Remove white membrane. Dissolve soda in water. Add rind to soda water. Cook 10 minutes. Add orange and lemon pulp (free from membrane). Cook 15 minutes more. Add berries and sugar. Bring to a boil. Simmer 5 minutes. Remove from heat and stir in pectin. Stir and skim 5 minutes. Pour into hot, sterilized containers. Cover with 1/8 inch paraffin. Creates four, six ounce glasses.

Blueberry Apple Butter

Submitted by: Stella Brown
From: Gardiner, ME

Ingredients:

6 to 8 Pounds apples   8 Cups blueberries   4 Cups sugar
1 Tbls. Allspice

Instructions:

Slice apples. Add water to cover. Cook until soft. Press through sieve or food mill. Measure eight cups apple pulp. Mix apple pulp, berries, sugar, and allspice. Cook until thick and smooth. Pour hot into sterile containers. Process 10 minutes in boiling water bath.

Molasses Blueberry Muffins

Submitted by: Alvina Menard
From: Topsham, ME
Bake Time: 20 minutes
Bake Temperature: 400 degrees

Ingredients:

1 Cup sugar   2 Tbls. molasses   3/4 Cup milk
1/3 Cup shortening   2 Cups flour   1/2 Tsp. nutmeg
1 egg   1/2 Tsp. salt   1 Cup blueberries
1/2 Tsp. soda

Instructions:

In a large mixing bowl cream together sugar and shortening. Add egg and molasses; beat well. Sift together flour, soda, salt, and nutmeg; add flour mixture and milk alternately to creamed mixture. Fold in blueberries. Place batter in greased muffin cups, filling 2/3 full. Bake at 400 degrees for 20 minutes or until nicely browned. These burn easily. Makes 15 muffins.

Creamy Blueberry Dip

Submitted by: Mrs. David Pearsall
From: Eliot, Maine

Ingredients:

2 Cups blueberries   1/3 Cup light cream cheese   1 Tbls. apricot preserves

Instructions:

In the container of a food processor or blender, place blueberries, cream cheese, and apricot preserves; whirl until smooth. Serve with sliced fruit or use as a dessert sauce spooned over cut fruit.

Blueberry Rhubarb Crumble

Submitted by: Marilyn Bean
From: Vienna, ME
Bake Time: 65 minutes
Bake Temperature: 350 degrees

Ingredients:

3 Cups blueberries   1/2 Cup sugar   1/2 Cup butter
1 1/2 Cup rolled oats   2 Tbls. Flour   2/3 Cup brown sugar
1/2 Cup flour   2 Cups rhubarb cut in 1-inch pieces

Instructions:

Thaw berries if frozen. Do not drain. Combine oats, brown sugar, and 1/2 cup flour. Cut in butter with pastry blender until mixture resembles course crumbs. Reserve 2/3 cup crumbs for topping. Pat remaining crumbs into the bottom of a greased 9 by 9 inch baking pan. Bake at 350 for 10 to 15 minutes. Combine berries and rhubarb. Add sugar and 2 tablespoons flour, toss to coat well. Spoon into baked crust. Sprinkle with reserved crumb mixture. Bake at 350 for 45 to 50 minutes or until golden. Serve warm with whipped cream. Makes 6 servings.

Blueberry Ice Cream

Submitted by: Gertrude Shibles
From: Knox, ME

Ingredients:

4 Cups blueberries   1/2 Tsp. vanilla   4 Cups heavy cream
juice of 1 lime   1/8 Tsp. salt
2/3 Cup honey (blueberry or light colored)

Instructions:

Puree blueberries in a blender. Frozen berries should be thawed. Add honey, limejuice, vanilla, and salt to the blueberry puree in blender. Mix. Stir puree mixture into cream and pour into a 4 quart freezer can of electric or hand cranked ice cream freezer. Freeze, using manufacturer's directions. Makes 2 1/2 quarts.

Blueberry Oat Bars

Submitted by: Edward King
From: Jay, ME
Bake Time: 30 minutes
Bake Temperature: 350 degrees

Ingredients:

1 3/4 Cups oatmeal
3 Tbls. water
2 Cups blueberries
1/2 Tsp. nutmeg

1/2 Cup chopped nuts
1 1/2 Cup flour
2 Tbls. cornstarch
1/2 Cup sugar

3/4 Cup butter
1/2 Tsp. baking soda
3/4 Cup brown sugar
2 Tsp. lemon juice

Instructions:

Combine oats, flour, brown sugar, nuts, baking soda, and nutmeg. Add butter, mixing until crumbly. Reserve 1-cup mixture; press the rest into bottom of greased 7 by 11 inch glass baking dish. Bake 10 minutes at 350. Combine blueberries, sugar, water, cornstarch, and lemon juice. Bring to a boil, reduce heat and cook 3 minutes, stirring occasionally, or until thickened. Spread over baked base. Bake 18 to 20 minutes or until topping is golden brown.

Peek-A-Blueberry farm

M ARYLAND

"You can tell this recipe was a favorite. The notations to the side are my mom's handwriting to make ½ more than the original recipe. We liked it so much , we never had enough with just the original amounts. Enjoy…"

Blueberry Cake

Submitted by:	Rosemarie Skovira
From:	Baltimore, MD
Bake Time:	35 minutes
Bake Temperature:	350 degrees

Ingredients:

1 1/2 Cups blueberries	1 Tsp. baking powder	1/3 Cup milk
1/2 Cup butter	2 eggs, separated	1 Tsp. vanilla
1 Cup sugar	1 1/2 Cups flour	1/4 Tsp. salt

Instructions:

Combine butter and sugar until light and fluffy. Add egg yolks and vanilla. Beat well. Sift dry ingredients. Add to butter mixture alternating with milk. Beat well after each addition. Fold in stiffly beaten egg whites. Spread half the batter in a greased and floured 9 inch square cake pan. Sprinkle berries across top of batter, cover with remaining batter. Bake at 350 for 35 minutes.

Blueberry Dumplings

Submitted by: Janet Paugh
From: Oakland, MD

Ingredients:

2 1/2 Cup fresh blueberries   2 Tsp. baking powder   1 Tbls. butter
1/3 Cup sugar   2 Tbls. sugar   1/2 Cup milk
1 Cup water   1 Tbls. lemon juice   1/4 Tsp. salt
1 Cup flour

Instructions:

Bring the blueberries, 1/3 cup sugar, salt, and water to a boil. Cover; simmer 5 minutes. Add lemon juice. Sift together dry ingredients; cut in butter until the mixture resembles a course meal. Add milk all at once; stir only until flour is dampened. Drop batter from tip of tablespoon into bubbling sauce, making 6 dumplings. Do not let them overlap. Cover tightly; cook over low heat 10 minutes without peeking. Serve hot.

Blueberry Cream Muffins

Submitted by: Lisa Swingholm
From: APG, MD
Bake Time: 20 minutes
Bake Temperature: 375 degrees

Ingredients:

4 Cups flour   2 eggs   1 egg
1 Cup sugar   2 Cups milk   (1) 8 Oz. cream cheese
6 Tsp. baking powder   2 Cups blueberries   1/2 Cup sugar
1 Tsp. salt   1/2 Cup butter

Instructions:

In a large bowl, combine flour, one-cup sugar, baking powder, and salt. In another bowl combine beaten eggs, milk, and butter. Stir into dry ingredients just until moistened. Fold in blueberries. Spoon about 2 round tablespoonfuls into a greased muffin pan. In a small bowl beat together cream cheese, egg, and 1/2 cup sugar. Place small teaspoonfuls onto batter already in muffin pan. Do not spread. Cover with another teaspoon of regular batter. Bake at 375 for 18 to 20 minutes. Cool in pan for about 10 minutes before placing on wire rack to cool. Makes about 2 dozen.

Blueberry and Chicken Pasta Salad

Submitted by: Chloe Smith
From: Baltimore, MD

Ingredients:

1/2 Cup finely chopped red pepper
1 Cup pea pods, cut in half
1/4 Cup red wine vinegar
1/2 Cup grated Parmesan cheese
1 Cup fat free red wine vinegar dressing

3 Cups spiral pasta
2 Cups cooked cubed chicken
2 Tbls. chopped fresh basil
1/4 Cup chopped red onion

1 Cup sliced celery
1 Cup fresh blueberries
Salt and pepper
1/4 Cup chopped parsley

Instructions:

Cook pasta according to directions on package. About 1 minute before it is cooked, add the pea pods. Drain and rinse with cold water. To a large bowl, add pasta and pea pods along with all the remaining ingredients except Parmesan cheese and red wine salad dressing. Toss the salad with 1/2 cup of the dressing. Cover; refrigerate several hours or overnight to blend flavors. Before serving, toss with the remaining dressing and cheese.

Peek-A-Blue berry farm

MASSACHUSETTS

"10 AM, 38 degrees, sunny and cold. I am sending you my copy of blueberry muffins that were very famous years ago. They were made by Jordan Marsh Company, which was a big department store in Boston. It is now known as Macy's so these muffins are no longer available. I am unable to do any baking as I had a stroke a year ago and am confined to a wheelchair. I am 94 years old and can still knit and crochet..."

Blueberry Gingerbread

Submitted by:	Beth Bickle
From:	Tewksbury, MA
Bake Time:	40 minutes
Bake Temperature:	350 degrees

Ingredients:

1/2 Cup butter
1 1/4 Cups blueberries
1/2 Tsp. salt
1 Tsp. ginger
1 Cup hot water

1 Tsp. baking soda
1/2 Cup brown sugar
Lemon sauce
1 egg

1/2 Tsp. cloves
2 1/2 Cups flour
1 Cup molasses
1 Tsp. cinnamon

Instructions:

Cream butter. Add sugar and molasses. Beat until light. Add egg and beat well. Sift dry ingredients and use a small amount to coat berries. Add remainder alternately with hot water to first mixture. Beat until smooth. Fold in floured berries. Bake in a greased 9 by 13 by 2-inch pan at 350 degrees until it's done, about 30 to 40 minutes. Serve with lemon sauce.

Jordan Marsh Department Blueberry Muffins

Submitted by: Shirley Bogdan
From: Saugus, MA
Bake Time   25 minutes
Bake Temperature: 375 degrees

Ingredients:

2 Cups flour
1/4 Tsp. cinnamon
1 Tbls. milk
3 Tsp. sugar
1/4 Tsp. nutmeg

2 Tsp. baking powder
1 1/4 Cup sugar
6 Tbls. butter
2 eggs

1 Pint blueberries
1 Tsp. vanilla
1/4 Cup milk
1/4 Tsp. salt

Instructions:

Cream sugar, butter, and vanilla well. Add eggs. Beat again. Alternate dry ingredients with milk. Add berries. Spoon into greased muffin tins or use paper liners. Mix the nutmeg, cinnamon, and sugar to create a topping. Sprinkle the top of the muffins with the topping. Bake at 375 for 25 minutes. Serve warm or cool on wire rack. Now Jordan Marsh department store is Macy's

Blueberry Apple Betty

Submitted by: Gloria Robinson
From: Holyoke, MA
Bake Time: 50 minutes
Bake Temperature: 350 degrees

Ingredients:

1 21 Oz. can apple pie filling
14 Oz. frozen blueberries
1 white cake mix

1 1/2 Cup chopped nuts
1/2 Cup butter

Ice cream
1 Cup sugar

Instructions:

Lightly butter a 9 by 13 inch baking pan and spread the apple pie mixture in it. Toss the blueberries with three quarter cup of the sugar and spoon it over the pie mix. Sprinkle the cake mix evenly over this and drizzle the melted butter on top. Sprinkle the rest with the chopped nuts and the rest of the sugar. Bake 45 to 50 minutes at 350 until golden and bubbling and serve hot with ice cream.

Blueberry Betty for 50

Submitted by: Gail Cauger
From: North Attleboro, MA
Bake Time: 45 minutes
Bake Temperature: 350 degrees

Ingredients:

5 Quarts stewed blueberries* =   (1 Cup of Water & 1 Cup of Sugar)
4 Quarts bread crumbs   2 Tsp. nutmeg   1 1/8 Cup fat
3 1/2 Cups brown sugar   3 Tsp. cinnamon   4 Tbls. lemon juice
3 1/2 Quarts blueberry juice or water

Instructions:

Cover bottom of greased pan with a layer of crumbs. Add layer of stewed blueberries, a layer of crumbs, and repeat until all are used. Have crumbs on top. Dissolve brown sugar in blueberry juice (or water). Add spices and lemon juice and pour over mixture. Melt fat and pour over top. Bake 350 degrees for 45 minutes. This can be served with vanilla icing, lemon icing, plain, or with whipped cream.

* Stewed Blueberries

Mix one cup sugar and one cup water. Bring to a boil and boil for 3 minutes. Add washed berries and let simmer 5 minutes. Creates about 5 servings.

Peek-A-Blue berry farm

MICHIGAN

"Blueberries"
Blueberries begging to be picked;
Nearly bursting with their goodness,
So round and plump they are that they
Make my mouth water with anticipation.
I chose a bush and pick the first berry.
Ahhh! Delicious is an understatement!
One berry devoured calls for another,
And another… then I remember my bucket.
To fill my bucket was why I came here!
I increase my will power to the top notch
 And cover the bucket's bottom
Before consuming another berry.
The fight continues; berry in the bucket vs. belly.
It is never-ending till both are heavy with berries.
Submitted by Vale Fay

"Hope you can read my writing, as I write with my left hand. My right side is crippled from a farm accident 13 years ago. I make hand stamped greeting cards to sell, assemble poem books to sell, and I also make wild bird feeders with plastic pop bottles. My sister and I worked in a big blueberry patch this summer. We have made this recipe a few times and it is delicious."

" We live in a blueberry growing area. Our neighbors have the largest blueberry farm in the county (Ottawa County) which produces the most berries for the state of Michigan. "

Blueberry Frosty

Submitted by: Ethan Knott
From: Boyne City, MI

Ingredients:

1 1/2 Cup blueberries
1 Cup frozen sliced peaches
1/3 Cup honey

1 Cup vanilla yogurt
1/2 Tsp. cinnamon

1/2 Tsp. nutmeg
1 Cup milk

Instructions:

Combine blueberries, peaches, and milk in a blender. Cover and process on high. Add yogurt, honey, cinnamon, and nutmeg; blend well. Pour into glasses. Garnish with cinnamon sticks if desired. Creates four 1-cup servings. Serve immediately.

Blueberries in the Snow

Submitted by: Connie Hamlin
From: Mc Millan, MI
Bake Time:
Bake Temperature:

Ingredients:

1/2 Cup sugar
1 1/2 Cups fresh blueberries
1 Can blueberry pie filling 21oz.

1 Carton whipped topping 16oz.
(1) 8 oz. Pkg. Cream cheese

1/2 Cup milk
1 large angle food cake

Instructions:

Combine sugar, milk, and cream cheese in a large bowl. Beat with electric mixer until blended. Fold in whipped topping and blueberries. Crumble angle food cake into small pieces and add to cream mixture. Mix well and pour into a large bowl, packing mixture down and spreading evenly. Pour blueberry pie filling on top and spread evenly. Cover and refrigerate for at least 3 hours before serving.

Blueberry Plum Cobbler

Submitted by: Katherine Easton
From: Grand Ledge, MI
Bake Time: 40 minutes
Bake Temperature: 375 degrees

Ingredients:

8 plums
1 Pint blueberries
1/2 Cup sugar
2 Tbls. flour

1/8 Tsp. salt
1 Cup flour
3/4 Cup baking powder
1/2 Cup buttermilk

1 egg white
1 1/2 Tbls. vegetable oil
4 Tsp. Sugar

Instructions:

Preheat the oven to 375 degrees. Coat an 8 by 8 inch baking dish with nonstick spray. In a large bowl, combine the plums, blueberries, 1/2 cup of sugar, and 2 tablespoons of flour. Pour into the prepared baking dish. In a medium bowl, combine the baking powder, salt and remaining 1 cup of flour, and 3 teaspoons of the remaining sugar. In a small bowl, combine the buttermilk, egg white, and oil. Add to the flour mixture. Stir until a thick batter forms. Drop the batter by tablespoon on top the fruits. Sprinkle with the remaining 1-teaspoon of sugar. Bake for 35 to 40 minutes, or until golden brown and bubbling. Transfer to a rack to cool. Serve warm or at room temperature. Makes 8 servings.

Blueberry Mustard Chicken

Submitted by: June Foster
From: Traverse City, MI

Ingredients:

4 Tbls. lemon juice
2 Tbls. vegetable oil
1 Cup diced onion

2 Tbls. Dijon mustard
2 Tbls. honey
2 Tsp. dry mustard

1/4 Tsp. cinnamon
2 Cups blueberries
4 Tbls. white wine

Instructions:

Marinate chicken in 2 tablespoon of the lemon juice and 2 tablespoons of oil for 3 to 4 hours. Sauté the diced onion in vegetable oil until soft; add the dry mustard, the rest of the lemon juice, Dijon mustard, honey, white wine, 2 cups blueberries and cinnamon. While sauce boils gently for 10 minutes, grill or cook chicken. Top with sauce and serve.

Purple Cow

Submitted by: Frances Williams
From: Manton, MI

Ingredients:

1 Cup milk   1/8 Tsp. vanilla   1 Cup blueberries
1 Cup vanilla ice cream

Instructions:

If blueberries are frozen, partially defrost before using. Reserve a few whole berries for garnish. Blend all ingredients until frothy and purple. For fizz add a splash of soda water. Top with whole berries. Make 2 1/2 Cups.

Blueberry Catsup

Submitted by: Gladys Sykes
From: Alger, MI

Ingredients:

2 Quarts blueberries   1 Tsp. mace   2 Tsp. ground cloves
1 Pint blueberries   3 Cups white vinegar   7 Cups sugar
1 Tbls. cinnamon

Instructions:

In a large saucepan, combine all ingredients and bring to a boil. Lower heat and simmer gently, stirring occasionally, about 2 hours or until mixture is thick and creamy. Pour mixture into sterilized jars. Seal and store in a cool dry place. Storage life on shelf, one year. Serve with chicken, ham, pork, duck, spareribs, hamburgers, hot dogs. Can also be spread on toast. A good glaze for kabobs to be brushed on during the last 15 minutes of cooking. Creates about 4 pints.

Michigan Blueberry Split Salad

Submitted by: Lottie Hathaway
From: Jonesville, MI

Ingredients:

1 Half Pint sour cream
1 Tbls. olive or salad oil
1 1/2 Cups blueberries
2 Tbls. wine vinegar

Bibb lettuce or chicory
1/4 Tsp. salt
2 Pounds creamed cottage cheese
3/4 Cup crumbled Danish blue cheese

1/8 Tsp. ground pepper
Lemon juice
6 bananas

Instructions:

Combine sour cream, Danish cheese, olive oil, wine vinegar, pepper, and salt. Chill. Arrange lettuce on six individual plates. Peel bananas and halve lengthwise. Coat all surfaces with lemon juice. Arrange two halves on each plate. Using ice cream scoop, place two scoops of cottage cheese on each plate. Garnish with blueberries. Serve with blue cheese dressing. Creates 6 servings.

Blueberry Breakfast Rolls

Submitted by: A Friend
From: Grand Junction, MI
Bake Time: 15 minutes
Bake Temperature: 375 degrees

Ingredients:

2 Tbls. sugar
2 Tbls. orange juice
1 Tsp. grated orange peel
1 10 Oz. can refrigerated pizza crust dough

3/4 Cups blueberries, chopped
1 Tbls. milk
2 Tsp. cornstarch

1/2 Cup powdered sugar
1/2 Tsp. grated orange peel

Instructions:

Preheat oven to 375. Coat 12 muffin cups with vegetable cooking spray. In a small saucepan, combine blueberries, orange juice, sugar, cornstarch, and 1 teaspoon grated orange peel. Stir to dissolve cornstarch. Cook over medium heat, stirring constantly until thick and bubbly (about 3 minutes). Set aside to cool for 10 minutes. Unroll pizza dough onto a lightly floured surface; pat into a 12 by 9 inch rectangle. Spread the blueberry mixture over dough, leaving a 1/2 inch boarder along the sides. Beginning with a long side, roll up jellyroll fashion; pinch seam to seal. Do not seal ends of roll. Cut roll into 12 one inch slices. Place slices; cut side up, in coated muffin cups. Bake 12 to 15 minutes or until lightly browned. Remove rolls from pan; cool on a wire rack for at least 15 minutes. Combine powdered sugar, milk, and 1/2 teaspoon grated orange peel, stirring until smooth. Drizzle icing over cooled rolls.

Blueberry and Tortellini Fruit Salad

Submitted by: Anonymous
From: Grand Junction, MI

Ingredients:
1/4 sliced almonds   1 Cup fresh blueberries   1/2 Cup poppy seed dressing
1 Cup sliced fresh strawberries 3/4 Cup green grapes
1 9 Oz. pkg. three cheese Tortellini pasta
(1) 11 Oz. can Mandarin orange segments

Instructions:
Cook pasta according to directions. Three cheese Tortellini pasta can be found in the refrigerated section of the grocery store. In a large bowl, add pasta, blueberries, strawberries, drained oranges, grapes, and almonds. Pour dressing over salad and toss lightly; refrigerate until ready to serve.

Blueberry Pumpkin Rum Pie

Submitted by: Gladys Sykes
From: Alger, MI
Bake Time: 50 minutes
Bake Temperature: 425 degrees

Ingredients:
3 eggs   1/2 Tsp. salt   1/2 Cup sugar
2 Tsp. rum flavoring   1/2 Tsp. ginger   1/4 Cup cold water
1/2 Cup sugar   1/2 Tsp. nutmeg   1 can blueberry pie filling
2 Cups blueberries   1/4 Tsp. Angostura aromatic bitters 1 Envelope unflavored gelatin
1/2 Cup milk   1 Tbls. rum flavoring
1 1/2 Cups cooked mashed pumpkin 1/2 Tsp. cinnamon 1 Cup dry-pack frozen blueberries
1 baked 9 inch pie shell, crumb crust type

Instructions:
In a saucepan, combine egg yolks, sugar, pumpkin, milk, salt and spices. Stir over low heat until mixture thickens. Mix together 2 teaspoons rum flavoring, Angostura bitters, gelatin and water. Stir mixture into hot pumpkin mixture until gelatin is dissolved. Chill until cold and slightly thickened. Beat egg whites until stiff. Gradually beat in sugar, 1 tablespoon at a time, until glossy. Fold egg whites and the 2 cups blueberries into pumpkin mixture. Pour mixture into baked pie shell. Chill until firm. In a bowl, mix the blueberry pie filling, one-cup blueberries, and tablespoon of rum flavoring. Spoon mixture around outer edge of pie. Chill until ready to serve. Creates one 9-inch pie.

MINNESOTA

"Greetings from Minnesota! My whole family loves when the local wild blueberries ripen so we can enjoy their wonderful flavor. This recipe was one of my mother's favorites and was gobbled up every time..."

"I love blueberries. We used to go and pick wild ones and I canned them- fun and work. It took all forenoon to pick a 5 quart pail full. They are much smaller than yours..."

Blueberry Boy Bait

Submitted by: Kathy Haug
From: Pine City, MN
Bake Time: 50 minutes
Bake Temperature: 350 degrees

Ingredients:

2 Cups flour
1 1/2 Cup sugar
2/3 Cup butter

2 Tsp. baking powder
1 Tsp. salt
2 eggs

1 Cup milk
3 Cups blueberries

Instructions:

Sift together the flour, sugar, and cut in the butter until the mixture resembles a coarse meal. Set aside 3/4 cup of the crumb mixture for topping. To the rest of the crumb mixture add baking powder, salt, egg yolks, and milk. Mix until well blended. Beat egg whites until stiff and fold into batter. Pour into a greased 8 by 12 pan. Arrange the blueberries over batter. Sprinkle remaining crumb mixture over the top. Bake at 350 for 40 to 50 minutes.

No Sugar Blueberry Banana Muffins

Submitted by: HelenBressler
From: Belle Plaine, MN
Bake Time: 15 minutes
Bake Temperature: 350 degrees

Ingredients:

2/3 Cup mashed banana	2 Cups flour	1 egg
1 Tsp. baking powder	1/2 Cup milk	1 Tsp. baking soda
1/3 Cup vegetable oil	1 Cup blueberries	

Instructions:

Beat together banana and egg until creamy. Add milk and oil; beat well. Mix in the flour, soda, and powder. Beat well. Gently mix in blueberries. Spoon batter into oiled and floured muffin tins. Bake at 350 for 15 minutes. Cool on wire racks.

Blueberry Muffins

Submitted by: Norma Wassenaar
From: Hills, MN
Bake Time: 25 minutes
Bake Temperature: 400 degrees

Ingredients:

1 egg	1 1/2 Cup flour	1/2 Tsp. salt
1/2 Cup milk	1/2 Cup sugar	1 Cup blueberries
1/4 Cup oil	2 Tsp. baking powder	

Instructions:

Beat together the egg, milk, and oil in a small bowl. Combine all the dry ingredients in a separate larger bowl. Add egg mixture, stirring with a fork until just moistened. Add blueberries. Bake at 400 in a greased muffin tin for 20 to 25 minutes.

Creamy Frozen Fruit Cups

Submitted by: Mrs. Harris
From: Fosston, MN

Ingredients:

1 8 Oz. pkg. cream cheese   2 Cups blueberries   1/2 Cup sugar
(1) 8 Oz. carton whipped topping

Instructions:

Beat the cream cheese and sugar until fluffy. Add the blueberries. Fold in whipped topping. Line muffin cups with foil liners. Freeze until firm. Remove from freezer 10 minutes before serving.

Blueberry Pineapple Bread

Submitted by: Sharon Fisher
From: Pine City, MN
Bake Time: 50 minutes
Bake Temperature: 350 degrees

Ingredients:

3 Cups flour   1 1/3 Cup sugar   1 1/2 Tsp. lemon juice
2 Tsp. baking powder   1/2 Tsp. salt   1/2 Cup milk
1 can crushed pineapple, drained   1 Tsp. soda   4 eggs
2 Cups blueberries   2/3 Cup shortening

Instructions:

Sift the flour, baking powder, soda, and salt together. Cream shortening until light and fluffy. Gradually beat in sugar. Stir in eggs, milk, lemon juice, and pineapple. Beat into the dry ingredients. Gently fold in blueberries. Pour into 6 greased and floured 6 by 3 by 2 pans. Bake at 350 for 40 to 50 minutes. Unmold and cool on rack.

Double Blueberry Pie

Submitted by: Linda Sinell
From: Arlington, MN
Bake Time:
Bake Temperature:

Ingredients:

(1) 9 inch baking crust   Whipped cream   (1) 16 Oz. jar blueberry jam
2 Cups fresh blueberries   1/4 Tsp. cinnamon

Instructions:

In a microwave safe dish, combine blueberry jam and cinnamon. Microwave on high until mixture liquefies, about 1 minute. Stir in fresh blueberries. Spoon into piecrust. Chill. Serve topped with whipped cream or ice cream. Creates 6 servings.

Blueberry Chicken Salad

Submitted by: Jeanne Grau
From: Bloomington, MN

Ingredients:

2 Cups cubed chicken   1/2 Cup mayo   2 Cup fresh blueberries
1/2 Tsp. ginger   2 Tsp. sugar   1/2 Cup sliced almonds
1 Tsp. fresh lemon juice   1 Cup thinly sliced celery 1/4 Cup sour cream
2 Tsp. Grated lemon peel   1 Cup green and red seedless grapes, halves

Instructions:

Combine the chicken, blueberries, grapes, celery, and almonds in a large bowl. Mix the mayo, sour cream, lemon juice, lemon peel, sugar, and ginger together. Pour over chicken mixture and gently stir.

Blueberry Raspberry Jam

Submitted by: Lyne Zebrasky
From: Wyoming, MN

Ingredients:

3 Cups sugar   1 3/4 Oz. pkg. Light powdered pectin
2 Pounds berries, about 2/3 blueberries to 1/3 raspberry

Instructions:

Puree fruit in a blender. Heat fruit and sugar to full boil and boil for 1 minute. Remove from heat, stir in pectin until well mixed. Skim off any foam. Pour into hot sterile jars. Be sure to get jam 1/8 inch of top of jar. Quickly cover with hot sterile lids and rings. Can process in water bath canner for 10

Peek-A-Blue berry farm

M ississippi

"My family also has a blueberry farm. We have 700 bushes. These are some recipes that go way back in our family so I hope you enjoy them…"

"Blueberry Explosion- Put 1 cup blueberries in microwaveable container. Cover with wax paper and microwave on high until berries explode -about 15 seconds. Pour on ice cream or pancakes!"

Blueberry Bundt Cake

Submitted by: Jo Pearl Odom
From: Petal, MS
Bake Time: 40 minutes
Bake Temperature: 350 degrees

Ingredients:

1 1/2 Cup fresh blueberries   1 Cup sour cream   1 lemon cake mix
4 eggs   2 Tbls. flour

Instructions:

Sprinkle flour over well-drained berries. Combine dry cake mix and sour cream. Add whole eggs, one at a time, while beating at slow speed. Beat at medium speed, scraping the bowl often. Fold in floured blueberries. Pour into cake pan. Bake at 350 degrees for 35 to 40 minutes. Cool in pan for 15 minutes before removing.

Blueberry Barbecue Sauce

Submitted by: Katherine Oglesby
From: Sardis, MS

Ingredients:

1 Tsp. olive oil
1/4 Cup chopped onions
1 jalapeno (seeded and chopped)
1 Pint fresh blueberries
2 Tbls. rice wine vinegar
1 1/2 Tbls. dark brown sugar
1 Tbls. Dijon mustard
1/4 Cup water

Instructions:

Sauté the onions and jalapeno in olive oil over medium high heat in a small saucepan until limp, 2 to 3 minutes. Add the remaining ingredients and cook at a low boil for 15 minutes, stirring often. Puree the sauce in a blender or food processor until smooth.

Grandpa's Blueberry Waffles

Submitted by: Arnold Pancratz
From: Jackson, MS

Ingredients:

1 Cup flour
2 Tsp. baking powder
3/4 Cup blueberries
1/2 Cup sugar
2 eggs
1 Tsp. salt
1/2 Cup peanut oil
1/2 Cup evaporated milk

Instructions:

Preheat waffle iron. While iron is heating, mash the berries and blend with other ingredients in a large bowl. Pour 1/4 cup batter into the waffle iron. When this is done, repeat as long as there is batter remaining. Serve with butter and syrup or honey. Makes about 6 waffles.

Blueberry Custard Pie

Submitted by: Daudra Preuss
From: Monticello, MS
Bake Time: 50 minutes
Bake Temperature: 400 degrees

Ingredients:

2 Cups blueberries
1 unbaked pastry pie shell
3 Tbls. flour

1 Cup sugar
2 eggs

1 stick butter, melted
1 Tsp. vanilla

Instructions:

Place blueberries in unbaked pie shell. Mix the sugar, eggs, flour, butter, and vanilla together. Pour over blueberries and bake at 400 degrees for 10 minutes; reduce heat to 350 and continue baking for 35 to 40 minutes.

Swedish Blueberry Grog

Submitted by: W.J. Morris
From: Carriere, MS

Ingredients:

1/2 Tsp. allspice
1 Tsp. whole cloves
1 Cup blueberry wine or sherry (optional)

2 cinnamon sticks
4 Cups blueberry juice, strained from extra ripe blueberries

3/4 Cup sugar

Instructions:

Tie cinnamon, cloves, and allspice in a spice bag. Add sugar and spice bag to juice. Simmer juice for 10 minutes. Remove spice bag. Add blueberry wine or sherry. Serve hot. Extra juice may be preserved in glass containers or frozen for future use.

Blueberry Ambrosia

Submitted by: W.J. Morris
From: Carriere, MS

Ingredients:

1 Cup blueberries
2 Cups diced marshmallows   1 can drained crushed pineapple   1 Cup sour cream
2 Tbls. sugar   1 can small green grapes or Emperor grapes halved
1 Cup peaches, cut in bite size pieces

Instructions:

Mix all ingredients together in a bowl. Let stand in refrigerator at least an hour, stirring occasionally. Creates 6 servings.

Honeyed Blueberry Sherbet

Submitted by: W.J. Morris
From: Carriere, MS

Ingredients:

3 Cups blueberries   1 Cup water   1/4 Cup orange juice
1/3 Cup sugar   1/3 Cup honey   Pinch of salt
1 Cup plain yogurt   Peel of 1 orange, cut into 3/4 inch squares

Instructions:

Combine blueberries, water, sugar, salt, and orange peel in a large sauce pan. Heat to boiling. Reduce heat and simmer 8 to 10 minutes. Remove orange peel and press mixture through a sieve. To 2 1/2 to 3 cups of puree, add honey and orange juice. Refrigerate at least 3 hours. Wisk yogurt into chilled blueberry mixture. Pour into an 8 inch metal pan; place in freezer. After 35 to 40 minutes mixture has frozen about an inch in from sides of pan, quickly scrape and fold frozen sides into center. Refreeze for about 20 minutes. Transfer lightly frozen mixture to a bowl and beat until aerated and lighter in color. Return to pan, freeze another 20 minutes and beat again. Scrape into storage container, seal lightly and return to freezer. Sherbet is ready when completely frozen.

Blueberry Pie

Submitted by: Janie Hyde
From: Booneville, MS
Bake Time: Varies
Bake Temperature: 350 degrees

Ingredients:

1 box white cake mix
1 can blueberry pie filling

1 Cup chopped pecans

1 stick butter

Instructions:

Place blueberries in a round pie dish. Put 1/2 of the dry cake mix. Place chopped pecans on top of the cake mix. Melt the butter and pour over top of the mixture. Bake at 350 until golden brown.

Blueberry Banana Pancakes

Submitted by: Ethel Howard
From: Lucedale, MS

Ingredients:

2 Cups sifted flour
1/4 Cup butter
1 Tsp. salt
2 Cups blueberries

1 1/2 Cups milk
2 Tsp. baking powder
2/3 Cup mashed banana

1/4 Cup sugar
1 Tsp. vanilla
2 eggs, well beaten

Instructions:

Mix together the flour, sugar, baking powder, and salt. Mix the eggs, milk, butter, banana, and vanilla. Add to the flour mixture. Fold in blueberries and cook on a hot greased griddle. Makes 6 to 8 servings.

Blueberry Lemon Cobbler

Submitted by: Carol Plowman
From: Steens, MS
Bake Time: Bake until golden
Bake Temperature: 375 degrees

Ingredients:

3 Cups blueberries   3 Tbls. butter   1/4 Tsp. vanilla
3/4 Cup sugar   1/4 Tsp. baking soda   1/4 Cup buttermilk
2 Tbls. grated lemon peel   1/4 Tsp. baking powder   3/4 Cup flour
1 egg   1/4 Tsp. salt   Whipping cream

Instructions:

Toss berries with 1/2-cup sugar and lemon peel in deep baking dish. Mix flour, remaining sugar, salt, baking powder and soda in a medium bowl. Cut in butter until crumbly. Beat egg, buttermilk and vanilla; add to flour mixture. Drop by spoonfuls over berries. Bake at 375 until topping is golden. Remove and cool slightly. Serve warm with cream. Creates 5 servings.

Missouri

"Here's a blueberry recipe from Missouri. We were first introduced to it at a potluck. Aren't potlucks great…"

"There's a blueberry farm just outside of town here and they give away recipes to people that have picked. I was born on Lincoln's birthday and on Lincoln Street here in town. I tell everyone they fly the flag on my birthday."

Blueberry Ice Cream

Submitted by: Sara Dent
From: Bevier, MO

Ingredients:

2 Cups heavy cream
1 Cup milk
1 14 Oz. can sweetened condensed milk
1 7 Oz. jar marshmallow cream
1 Pint blueberries
1 Tbls. vanilla

Instructions:

Gradually add condensed milk to marshmallow cream, beating until well blended. Stir in heavy cream, milk, and vanilla. Pour mixture into a 13 by 9 inch pan, cover, freeze until almost solid but still mushy in center. Scrape mixture into a large chilled bowl, beat with electric mixer until smooth. Repeat freezing and beating process once. Crush blueberries to measure 1 1/2 cup. Beat into ice cream mixture and refreeze until firm.

Baked Blueberry Sandwiches

Submitted by: Sara Dent
From: Bevier, MO
Bake Time: 20 minutes
Bake Temperature: 350 degrees

Ingredients:

3 Cups fresh blueberries   1 can sweetened condensed milk   1/3 Cup sugar
1 can flaked coconut   1/2 Cup water
12 slices white bread, crusts trimmed

Instructions:

Brush bread slices on both sides with condensed milk. Dip slices into coconut. Put 6 slices on a heavily greased cookie sheet. Cover slices with 2 cups blueberries, top with remaining slices. Bake at 350 for 15 to 20 minutes, or until lightly browned. Combine remaining blueberries, water, and sugar in a saucepan. Cook over moderate heat until berries are cooked and sauce is thickens. Serve warm with blueberry sauce spooned over each sandwich.

Spicy Apple Blueberry Crunch

Submitted by: Sandra Hughes
From: Hollister, MO

Ingredients:

5 Cups apples, peeled and sliced   1/4 Cup butter   3/4 Cup blueberries
1/4 Tsp. ground cinnamon   1/2 Cup flour   1/2 Cup chopped pecans
1/2 Cup brown sugar   2/3 Cup rolled oats   1/4 Tsp. nutmeg
(1) 4-serving size regular butterscotch pudding mix

Instructions:

Place apples and blueberries in a 8 by 1 1/2 inch round microwave safe dish. Sprinkle with 2 tablespoons of pudding mix. Set aside. In a medium bowl, melt butter, covered on high 30 to 45 seconds. Add remaining pudding, oats, flour, brown sugar, pecans, and spices. Stir until crumbly. Spoon over fruit mixture. Cook, uncovered on high 11 to 13 minutes- giving the dish a quarter turn every 3 minutes. Let stand 30 minutes before serving.

Blueberry Pumpkin Streusel Muffins

Submitted by: Jo Ann Lanius
From: Sullivan, MO
Bake Time: 30 minutes
Bake Temperature: 350 degrees

Ingredients:

2 1/2 Cups flour
2 Cups sugar
1 Tbls. pumpkin pie spice
1 Tsp. baking soda
1/2 Tsp. cinnamon

2 eggs
1/4 Cup vegetable oil
1 Cup blueberries
1 1/4 Cup pumpkin

1/4 Cup sugar
2 Tbls. flour
2 Tbls. butter
1/2 Tsp. salt

Instructions:

Preheat oven to 350. Grease or paper line 24 muffin cups. Combine 2 1/2 cups flour, 2 cups sugar, pumpkin pie spice, baking soda, and salt in a large bowl. Combine pumpkin, eggs and vegetable oil in a medium bowl; stir well. Stir into flour mixture just until moistened. Gently fold in blueberries. Spoon batter into prepared muffin cups, filling 3/4 full. Mix the 1/4 cup sugar, 2 tablespoons flour, and cinnamon in a medium bowl. Cut 2 tablespoons butter with pastry blender until crumbly. Sprinkle over the full muffin cups. Bake until golden on top.

Blueberry Raspberry Tart

Submitted by: Lucille Graff
From: Cuba, MO

Ingredients:

2 Tbls. orange flavored liqueur
1 10 inch prepared sponge flan cake
1/3 Cup current jelly, melted

2 Cups fresh blueberries
2 Cups frozen vanilla yogurt

1 Cup fresh raspberries

Instructions:

Brush liqueur evenly over cake; brush with 2 tablespoons of the melted jelly. Spoon the blueberries and raspberries into cake; brush berries with remaining jelly. Just before serving, spoon frozen yogurt into large bowl; beat at high speed until fluffy. Cut cake into wedges; top each wedge with a dollop of whipped yogurt. Makes 8 to 10 servings.

Blueberry Amaretto Squares

Submitted by: Norma Reed
From: Washburn, MO
Bake Time: 30 minutes
Bake Temperature: 350 degrees

Ingredients:

2 eggs   3/4 Cup sugar   5 Cups blueberries
1/2 Cup sugar   1/4 Cup cornstarch   1/2 Cup butter, melted
1/2 Cup amaretto divided   1 1/4 Cup sugar   2 1/2 Cups milk
1 8 oz. pkg. cream cheese   (2) 3 3/4 Oz. vanilla instant pudding mix
1 8 oz. whipped topping   1 3/4 Cups graham cracker crumbs

Instructions:

Combine graham cracker crumbs, 1/2 cup sugar, and butter. Press into a greased 13 by 9 inch pan. Combine eggs, 3/4 cup sugar, and cream cheese. Beat with mixer until smooth. Spread mixture over graham cracker crumbs. Bake at 350 for 30 minutes then cool completely. Combine milk, pudding mix, and 1/4 cup amaretto. Beat 2 minutes at low speed. Spread over cream cheese layer. Combine 1 1/4 cups sugar and cornstarch in a large saucepan. Gradually add 1/4 cup amaretto, stirring until smooth. Stir in blueberries. Cook over medium heat, stirring constantly until thickened, then cool. Pour blueberry mixture over pudding mixture. Chill thoroughly. To serve spread a 8 ounce tub of whipped topping over blueberries.

Lemon Blueberry Poppy Seed Bread

Submitted by: Virginia Owens
From: Springfield, MO
Bake Time: 62 minutes
Bake Temperature: 350 degrees

Ingredients:

2 Tbls. poppy seeds   1 egg   3/4 Cup water
1/2 Cup confectioners' sugar   1 Tbls. grated lemon peel   1 Tbls. lemon juice
1 pkg. Duncan Hines Bakery Style Blueberry with Crumb Topping Muffin Mix

Instructions:

Preheat oven to 350. Grease and flour an 8 by 4 inch loaf pan. Rinse blueberries with cold water and drain. For a loaf, empty muffin mix into medium bowl. Add poppy seeds, stir to combine and break up any lumps. Add egg and water; stir until moistened, about 50 strokes. Fold in blueberries and lemon peel. Pour into pan. Sprinkle topping from packet over batter. Bake at 350 for 57 to 62 minutes or until toothpick inserted in center comes out clean. Cool in pan for 10 minutes. Loosen loaf from pan. Lay foil overtop when removing from pan to keep topping intact. Invert onto cooling rack. Turn right side up. Cool completely. For a drizzle glaze, combine confectioners' sugar and lemon juice in a small bowl. Stir until smooth. Drizzle over loaf.

Blueberry Honey Syrup

Submitted by: Elenora Selenke
From: Shell Knob, MO

Ingredients:

4 Cups blueberries   3 Tbls. cornstarch   3/4 Cup water
1 Cup honey   1/4 Tsp. salt   1/4 Cup lemon juice

Instructions:

Mix together blueberries and honey. Add just enough water to cover the berries. Boil 5 minutes. Stir in cornstarch, salt, and 3/4 cup water. Cook until thickened. Add lemon juice. Serve hot or cold.

Blueberry Bread Pudding with Spiced Sauce

Submitted by: Mary Garrison
From: Saint Louis, MO
Bake Time: 70 minutes
Bake Temperature: 350 degrees

Ingredients:

8 eggs   1/4 Tsp. Salt   1 Tsp. cinnamon
2 Cups heavy cream   1 Tbls. butter   2 Cups milk
1/2 Tsp. ground cardamom   1/4 Tsp. ground nutmeg   3 Cup sugar
3 Tbls. lemon juice   1 1/2 Tbls. vanilla   5 Cups blueberries
1/4 Tsp. ground nutmeg   1 stick unsalted butter
12 lightly packed cups Italian bread, 3/4 inch cubes

Instructions:

In a large bowl, wisk eggs, cream, milk, 1 cup sugar, vanilla, 1/4 teaspoon nutmeg, and salt. Butter shallow 3 quart baking dish with 1 tablespoon butter. In a large nonstick skillet, melt half the remaining butter over medium high heat; add half the bread cubes. Sauté 8 minutes, tossing and stirring frequently, until lightly browned; add to cream mixture in bowl. In remaining butter, sauté remaining bread as directed; add to bowl along with 2 cups blueberries. Stir until bread is evenly moistened; pour into prepared baking dish. Cover; refrigerate overnight. In a food processor, process 3 cups blueberries until pureed; pour into medium saucepan. Stir into sugar; bring to a boil. Boil over medium heat 5 minutes. Scrape mixture through fine mesh sieve into 1-quart measure or bowl. Stir in lemon juice, cinnamon, cardamom, and 1/4-teaspoon nutmeg. Let cool; refrigerate for up to 3 days. Preheat oven to 350. Bring 8 cups of water to a boil. Uncover bread mixture. Place baking dish in a large roasting pan, pour enough boiling water into roasting pan to come up halfway up sides of baking dish. Bake 1 hour and 10 minutes, covering top of pudding with foil if browned too quickly, until knife inserted in center comes out slightly wet. Let cool on wire rack until pudding is warm. Reheat blueberry sauce in microwave until warm. To serve, dust top of pudding with confectioners' sugar. Serve with blueberry sauce.

Peek-A-Blue berry farm

Montana

"We truly enjoy Montana. The big cities are a bit overwhelming to us. Our mountains are so majestic…"

"We live in the middle of Montana on the Yellowstone river. "The Horse Whisperer" and "Frontier House" were movies filmed here…"

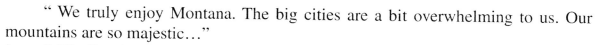

Blueberry Green Grape Dessert

Submitted by: Arlene Jones
From: Eureka, MT

Ingredients:

2 Tbls. vanilla
1 1/4 Quarts sour cream

2 Cups blueberries
2 1/2 Pounds seedless green grapes

1-Pound brown sugar

Instructions:

Mix grapes and blueberries with sour cream and vanilla. Place in baking dish; sprinkle brown sugar on top. Set dish in a pan of crushed ice and place under broiler. Cook until sugar is caramelized. Cool, then chill. The same ingredients can be used uncooked and served in parfait glasses as pretty, cool summer desserts. Serves 12.

Rhubarb and Blueberry Jam

Submitted by: Albertina Kaul
From: Great Falls, MT

Ingredients:

5 Cups sugar
(1) 6 Oz. pkg. raspberry Jell-o

1/8 Tsp. soda
1 Cup water

5 Cups diced rhubarb
1 Can blueberry pie filling

Instructions:

Bring the sugar, rhubarb, and water to a boil. Add soda. Stir until foam is gone. Add blueberry pie filling, boil 7 minutes. Remove from heat. Add Jell-O. Skim and cool jam. Put in jelly jars. May freeze for several months.

Blueberry Mallow Pie

Submitted by: Holly Diedrich
From: Shephard, MT

Ingredients:

2 1/2 Cup blueberry pie filling
1/2 Cup heavy cream, whipped
1 9 inch graham cracker crust, chilled

1/4 Cup milk
2 Cups mini marshmallows

1/2 Tsp. vanilla

Instructions:

Pour pie filling in crust. Chill. Melt marshmallows with milk a few seconds at a time in the microwave or in a double boiler. Stir until smooth. Add vanilla. Chill until slightly thickened. Mix until well blended. Fold in whipped heavy cream. Spread over filling. Chill until firm.

Blueberry Pie Filling

Submitted by: Kristi Kissner Schumacher
From: Great Falls, MT

Ingredients:

12 Cups blueberries   1 Tbls. grated lemon peel   3/4 Cup cornstarch
1/4 Cup lemon juice   3 Cups sugar

Instructions:

Wash and drain the blueberries. Combine sugar and cornstarch. Stir in blueberries; let stand until juice begins to flow, about 30 minutes. Add lemon peel and juice. Cook over medium heat until mixture begins to thicken. Ladle pie filling into can or freezer jars or plastic freeze bags, leaving 1/2-inch headspace. Cool at room temperature. Do not exceed 2 hours. Seal, label, and freeze. Creates 5 pints.

Peek-A-Blue berry farm

NEBRASKA

"My husband has a sister that lives in Michigan. He went to visit and pick some blueberries for us. The first time, he bundled them into plastic bags and packed them into his suitcase!! He was strongly encouraged (by me) to find another way to transport them. He now packs them in boxes and brings them home as luggage. This last trip it was 70 pounds…."

" We live in the North West corner of Nebraska, near the Wyoming border. We are about 30 miles from Crawford, where the mammoth bones were found in 1962.

Frozen Blueberry Dessert

Submitted by: Stacy Namuth
From: Lewellen, NE

Ingredients:
1 12Oz. tub Cool Whip   1 can blueberry pie filling
1 small can crushed pineapple (drained)
1 can Eagle Brand Milk (sweetened) 1 Cup mini marshmallows (optional)

Instructions:
Mix all ingredients together in a 9 by 13 inch cake pan or glass casserole dish. Freeze and serve.

Blueberry Drop Cookies

Submitted by: Loretta Heathers
From: Curtis, NE
Bake Time: 10 minutes
Bake Temperature: 375 degrees

Ingredients:

3/4 Cup sugar   1 Tsp. cinnamon   1 Cup blueberries
2/3 Cup shortening   1 egg   1 1/2 Cup sifted flour
1 Tsp. vanilla   4 Tsp. milk   1 1/2 Tsp. baking powder

Instructions:

Cream sugar, shortening, vanilla, and cinnamon together. Beat in eggs; stir in milk. Sift together flour, baking powder. Add to creamed mixture, mixing well. Fold in blueberries. Drop by teaspoonful on top greased cookie sheet. Bake at 375 for 8 to 10 minutes. Cool on rack.

Blueberries in the Snow

Submitted by: Edith Jelinek
From: Kearney, NE
Bake Time: 25 minutes
Bake Temperature: 350 degrees

Ingredients:

6 egg whites   2 Cups soda crackers   (1) 8 Oz. pkg. Cool Whip
3/4 Tsp. cream of tartar   3/4 Cup chopped nuts   1 Qt. Blue Berries
(1) 8 Oz. pkg. cream cheese 1 Tsp. lemon juice   2 Cups sugar
2 Tsp. vanilla   1 Cup sugar   3 Tbls. cornstarch

Instructions:

Beat eggs until frothy; add cream of tartar, beat until stiff and gradually add sugar. Mix well. Add crackers, chopped nuts, and vanilla. Spread in a buttered pan and bake at 350 for 25 minutes. When it cools it will sink. Mix cream cheese, softened with Cool Whip and spread over crust. Create a topping by heating 1 quart blueberries, 1 teaspoon lemon juice, 1 cup sugar, and 3 tablespoons cornstarch in a saucepan. Cook over low until thick. Cool and spread over Cool Whip mixture. Keep refrigerated.

Flat Blueberry Pie

Submitted by: Barbara Soseman
From: Omaha, NE
Bake Time: 45 minutes
Bake Temperature: 425 degrees

Ingredients:

1 unbaked Pie Crust	1/2 Cup flour	2 Tbls. lemon juice
Cornflakes	3/4 Tsp. cinnamon	3 Tbls. butter
1 1/2 Cups sugar	6 Cups blueberries	1 egg white

Instructions:

Line a 15 by 10 pan with piecrust, leaving some to fold over, around edge. Crush some corn flakes or other crunchy cereal and sprinkle over the bottom of the crust. This helps the crust not get so soggy. Mix the sugar, flour, cinnamon, berries, and lemon juice together gently. Spread over the bottom crust. Dot top of fruit mixture with butter. Roll out and lay top crust over fruit. Fold bottom crust around edges and seal as much as possible. This will often look patched on top (it is) but once finished looks wonderful. Lightly whip an egg white and brush over entire top crust, especially edges. Sprinkle slivered almonds over top before baking (optional). Bake at 425 for 35 to 45 minutes or until nicely browned like a pie. To serve, drizzle a powdered sugar glaze overtop.

Blueberry Lemon Drop Cookies

Submitted by: Melissa Arnold
From: Omaha, NE
Bake Time: 13 minutes
Bake Temperature: 375 degrees

Ingredients:

2 Cups flour	1/2 Cup butter	1 Tsp. baking powder
1 Tbls. flour	1 egg	1/2 Tsp. baking soda
1 Cup fresh blueberries	1 Tsp. grated lemon peel	1/2 Tsp. salt
1 Cup sugar	1 Tbls. lemon juice	

Instructions:

Heat oven to 375. Lightly grease cookie sheets. Set aside. In a small mixing bowl, combine 1 tablespoon flour and the blueberries. Toss to coat. Set aside. In a large mixing bowl, combine sugar, butter, egg, lemon peel and juice. Beat at medium speed until light and fluffy. Add remaining 2 cups of flour, baking powder, baking soda and salt. Beat at low speed until soft dough forms. Gently fold in blueberries. Drop dough by heaping teaspoons 2 inches apart onto prepared cookie sheets. Bake for 11 to 13 minutes, or until edges are golden brown. Cool completely before storing.

Blueberry Treat

Submitted by: Pat Cook
From: Wayne, NE
Bake Time: 20 minutes
Bake Temperature: 375 degrees

Ingredients:

2 eggs
1/4 Cup sifted confectioners' sugar
1/4 Cup melted butter
1/4 Tsp. grated orange rind
1 1/3 Cup vanilla wafer crumbs

1/3 Cup sugar
1/4 Tsp. salt
2 Cups blueberries
(1) 8 Oz. pkg. cream cheese

2 Tbls. cornstarch
1/2 Cup sugar
3/4 Cup water

Instructions:

Mix together crumbs, confectioners' sugar, and butter. Press in layer 8 by 8 by 2 inch pan. Beat eggs, 1/3 cup sugar, cream cheese and salt. Blend together. Pour over crumb crust. Bake at 375 for 20 minutes. Cool. Cover with 1 cup blueberries. Blend together 1/2 cup sugar, cornstarch, salt, and water. Add remaining blueberries and rind. Cook over low heat until clear and thick, stirring constantly. Pour over berries. Chill. May garnish with whipped cream.

NEVADA

" You just happened to pick my favorite fruit, and of course everyone knows my love of cooking and eating. If you get to make this recipe I guarantee it will be a hit…"

Blueberry Jam

Submitted by: Margaret Stephens
From: Fallon, NV

Ingredients:

2 1/2 Cups blueberries, mashed
3/4 Cup water

3 1/2 Cups sugar
1/2 Cup light corn syrup

1 Tbls. lemon juice
2 Oz. powdered pectin

Instructions:

In a large bowl, combine blueberries and lemon juice and then add corn syrup. Stirring constantly, gradually add sugar and let sit 10 minutes. In a 1 quart saucepan, combine pectin and water, bring to a boil. Boil for one minute, stirring constantly. Add pectin to blueberries, stirring to dissolve sugar. Pour jam into clean jars and seal. Let sit at room temperature for 24 hours before refrigerating or freezing. Makes six half pint jars.

Blueberry Pie Deluxe

Submitted by: Annette Morse
From: Las Vegas, NV

Ingredients:

4 Cups fresh blueberries   1 baked 9 inch pie crust   1 Cup sugar
Juice of half a lemon   3 Tbls. Cornstarch   1 Cup heavy sweetened
Cream- whipped

Instructions:

Boil 1-cup blueberries with 1 cup water and sugar, strain, reserving liquid. Add cornstarch to reserve liquid; boil in saucepan, stirring until thickened. Place remaining blueberries in a large bowl; add lemon juice. Pour boiling mixture over blueberries, cool. Place whipped cream in pie shell. Spoon blueberry mixture over cream. Chill several hours.

Blueberry Bonnets

Submitted by: Edith Peterson
From: Elko, NV
Bake Time: 28 minutes
Bake Temperature: 400 degrees

Ingredients:

2 sticks butter   2 Cups flour   1/2 Tsp. almond extracts
1 Tbls. confectioners' sugar   Sliced almonds (about 40 slices)
1/2 Cup blueberry preserves   1/2 Cup finely ground blanched almonds
1/2 Cup confectioners' sugar

Instructions:

Preheat oven to 400. In a large bowl, combine butter, 1/2 cup powdered sugar, and almond extract. Beat on medium speed until light and fluffy. Scrape down sides of bowl. Add flour and ground almonds and mix just until dough comes together and is blended. Using level tablespoon of dough, form 20 1 1/4 inch balls. Place dough balls, two inches apart on ungreased cookie sheets. Using the bottom of a small, flat measuring cup, flatten each ball into 2 inch circles 1/4 inch thick. Dip the bottom of the measuring cup into flour between each pressing to keep it from sticking to dough. Repeat with remaining dough balls. Bake for 7 to 8 minutes or until set and lightly golden around the edges. Remove cookies to wire racks to cool. Reduce oven to 325 degrees. Using rounded teaspoonfuls of dough, shape remaining dough into 20 1 inch balls. Place on ungreased cookie sheets, 1 inch apart. With fingertips, press balls to flatten slightly. Bake 15 to 20 minutes or until set and lightly golden around the edges. Remove and place on a wire rack to cool. Spoon one-teaspoon blueberry preserves on center of each round, flat cookie. Top with ball-shaped cookie, pressing lightly to allow preserves to show. Place two sliced almonds into preserves on each cookie and dust with remaining 1-tablespoon confectioners' sugar. with ball-shaped cookie, pressing lightly to allow preserves to show. Place two sliced almonds into preserves on each cookie and dust with remaining 1-tablespoon confectioners' sugar.

NEW HAMPSHIRE

"I'm 76 and live in a small New England town that was settled in the 1600's. I have 5 bushes that provide me with many berries. When I was a child we picked low bush ones, but mine are 4 to 7 feet tall. I cover them with black netting so the birds don't get them all."

Blueberry Butterscotch Russe

Submitted by: Mrs. Lester Stevens
From: Sharon, NH

Ingredients:

1 Cup heavy cream, whipped	2 Cups milk	Lady Fingers
2 Tsp. Angostura aromatic bitters	2 Cups blueberries	2 Tsp. grated orange rind
1 pkg. butterscotch pudding and pie filling		

Instructions:

Prepare pudding according to package directions, using two cups milk. Cook while stirring until mixture bubbles and thickens. Remove from heat, cover and cool. Fold in Angostura, orange rind, blueberries and whipped cream. Line sherbet glasses with lady fingers. Spoon blueberry mixture into center and top with additional blueberries. Chill until ready to serve. Creates 6 servings.

Blueberry Fungi

Submitted by: Terry Davis
From: W. Springfield, NH

Ingredients:

1 Quart blueberries
1/2 Cup sugar
1 Cup water

2 Cups flour
4 Tsp. baking powder
1 Tsp. sugar

1 Tbls. butter
1/3 Cup milk
1 Tbs. shortening

Instructions:

Mix blueberries, 1/2 cup sugar, and water in a small saucepan. Cover and boil gently until there is plenty of juice. Turn to medium, then to low. Combine flour, baking powder, and 1 teaspoon sugar.. Cut in butter and shortening. Add enough milk to make soft dough. Drop tablespoons onto hot blueberry mixture. Cover and do not peek for 15 minutes. Serve hot with vanilla ice cream or whipped cream.

Blueberry Bread

Submitted by: Dorothy Shaw
From: Franklin, NH
Bake Time: 50 minutes
Bake Temperature: 350 degrees

Ingredients:

1 egg
1 Cup milk

2 Tsp. baking powder
2 1/2 Cups flour

1/2 Cup sugar
1 1/2 Cups blueberries

Instructions:

Mix dry ingredients; add berries. Beat egg and milk, add to dry ingredients. Put mixture into greased and sugared loaf pan. Sprinkle top with sugar and bake for 50 minutes at 350.

Blueberry Batter Cake

Submitted by: Gladys McEwan
From: Errol, NH
Bake Time: 60 minutes
Bake Temperature: 375 degrees

Ingredients:

2 Cups blueberries
1 1/2 Tbls. lemon juice
3/4 Cup sugar
3 Tbls. butter

1 Cup sifted flour
1/2 Cup milk
1 Tsp. baking powder
1/2 Tsp. salt

1 Cup sugar
1 Cup boiling water
1 Tbls. cornstarch

Instructions:

Line a well greased 8 by 8 by 2 inch pan with berries, sprinkled with lemon juice. Cream 3/4 cup sugar and butter. Add milk alternately with flour, baking powder and 1/4 teaspoon salt which have been sifted together. Spread batter evenly over berries. Combine 1 cup sugar, 1/4 teaspoon salt and cornstarch. Sprinkle over top of cake. Pour boiling water all over. Bake at 375 for an hour. Creates 6 to 8 servings. Serve with whipped cream or ice cream.

Blueberry Ricotta Squares

Submitted by: Cindy Ames
From: Hinsdale, NH
Bake Time: 60 minutes
Bake Temperature: 350 degrees

Ingredients:

1 Cup flour
3/4 Cup sugar
1 1/4 Tsp. baking powder
1/4 Tsp. salt

1/3 Cup milk
1 1/2 Cup blueberries
1/2 Tsp. vanilla
1/4 Cup shortening

3 eggs
1/4 Tsp. vanilla
1/3 Cup sugar
1 1/4 Cup Ricotta cheese

Instructions:

In a small mixer bowl, combine the flour, 3/4 cup sugar, baking powder and salt. Add milk, shortening, 1 egg, and the 1/2 teaspoon vanilla. Beat with electric mixer on low speed until combined. Beat on medium for one minute. Pour batter into a greased 9 by 9 by 2 inch baking pan. Spread evenly. Sprinkle blueberries over batter. In another bowl, lightly beat the 2 eggs with a fork. Add the cheese, 1/3 cup sugar, 1/4 teaspoon vanilla, beat until combined. Spoon Ricotta mixture over blueberries and spread evenly. Bake at 350 for 55 to 60 minutes or until a inserted knife near the center comes out clean. Cool. Cut evenly into squares. Store bars, covered, in the refrigerator.

Peek-A-Blue berry farm

Nᴇᴡ Jᴇʀsᴇʏ

"The high bush blueberry was first cultivated in the early 1900's when Elizabeth White of Whitesbog, New Jersey began collecting blueberry bushes that grew the largest fruit.

Blueberry Sorbet

Submitted by: William Tindall
From: Trenton, NJ

Ingredients:

4 Cups blueberries   1/2 Cup sugar   1/4 Cup water
2 Tbls. lemon juice

Instructions:

Place blueberries and water in saucepan. Cook over medium high until blueberries are soft, stirring often. Press through strainer reserving juice, discard skin and seeds. Combine blueberry juice and sugar, stirring to dissolve. Add lemon juice. Cool. Pour into ice cream maker and freeze according to manufacturer's directions. Serve garnished with mint. Makes 2 1/2 cups to 3 cups sorbet.

Blueberry Foster

Submitted by: Lynn Ebeling
From: Hopewell, NJ

Ingredients:

2 Tbls. butter
3 Tbls. blackberry or raspberry jam

3/4 Tsp. cinnamon
1 large ripe banana

2 Cup blueberries
2 Tbls. sugar

Instructions:

In a skillet melt butter. Mash the banana and add to skillet along with sugar and cinnamon. Cook. Stir until sugar dissolves and then stir in jam. Add blueberries and cook 5 minutes. Serve over vanilla ice cream.

Blueberry Gingerbread

Submitted by: Agnes Vincent
From: Livingston, NJ
Bake Time: 45 minutes
Bake Temperature: 350 degrees

Ingredients:

2 Cups blueberries
1/2 Tsp. salt
3 Tbls. molasses
1 Tsp. soda
1 Cup margarine

1 egg
1 Tbls. sugar
Sugar
1 Tsp. cinnamon
1 Cup buttermilk

2 Cups flour
1 Cup sugar
2 Tbls. flour
Cinnamon
1/2 Tsp. ginger

Instructions:

Drench blueberries with the heaping tablespoons of sugar and flour, and set aside. Beat egg, one-cup sugar, margarine and molasses. Add buttermilk. Sift together 2 cups flour, ginger, salt, soda, and 1 teaspoon cinnamon. Add to egg mixture and beat 3 minutes. Fold in half the berries and pour into a 9 by 13 inch greased and floured pan. Add remaining berries. Pat down. Sprinkle top with a mixture of 3 heaping tablespoons of sugar and 1 teaspoon cinnamon. Bake 45 minutes at 350. Serve with whipped cream.

Happiness Pie

Submitted by: Katherine Gibson
From: Wayne, NJ

Ingredients:

1 Pint blueberries	20-25 strawberries	2 Cups milk
Whipped Topping	3 Oz. pkg. cream cheese	1/2 Tsp. vanilla

(1) 8 inch prepared graham cracker crust
1 small pkg. cooking and serve vanilla pudding mix

Instructions:

Combine pudding mix and milk in a saucepan. Bring to a rolling bubble over medium heat, stirring constantly. Remove from heat. Add cream cheese and stir until smooth. Add vanilla extract. Let mixture cool for 5 minutes, stirring twice. Pour pudding mixture into graham cracker crust. Refrigerate overnight or a minimum of 3 hours. Place strawberries in a circle around edge of pie and place one strawberry in center of pie. Place blueberries over remaining pudding surface. Serve with a whipped topping or whipped cream.

Peek-A-Blue berry farm

NEW MEXICO

"It's good that you started the collection at a young age. I started my recipe collection at 17 and I am 60 now. I live in Alburquerque- the true spelling. Years ago it was shortened by an r. I like the true spelling myself…"

Shiny-Top Cobbler

Submitted by: Sally Morge
From: Las Cruces, NM
Bake Time: 60 minutes
Bake Temperature: 350 degrees

Ingredients:

5 Cups blueberries
1 1/2 Tbls. lemon juice
2 Cups flour
2 Tsp. baking powder

1/3 Cup margarine
1 Cup milk
1 Tsp. salt

2 Tbls. cornstarch
1 1/2 Cup boiling water
3 Cups sugar

Instructions:

Spread berries in a well greased 9 by 13 shallow baking dish. Sprinkle with lemon juice. Set aside. In a medium bowl, stir flour, 1 1/2 cup sugar, milk, margarine, baking powder, and 1/2 teaspoon salt. Mix until well blended. Batter will be thick. Spoon over berries. Spread to edges of baking dish and set aside. In a small bowl, mix the remaining 1 1/2 cup sugar, cornstarch, and the remaining 1/2 teaspoon salt. Sprinkle over the batter. Pour boiling water carefully over all. Bake at 350 for an hour or until golden brown and glazed. Serve warm or cooled with whipped cream or ice cream.

Lemon Blueberry Salad

Submitted by: M. Berg
From: Albuquerque, NM

Ingredients:
(1) 3 Oz. pkg. lemon Jell-o   1 Tsp. lemon juice   1 Cup boiling water
1 3 Oz. pkg. blackberry Jell-o   1 Cup confectioners' sugar, sifted 1 Cup sour cream
(1) 21 Oz. can blueberry pie filling

Instructions:
Dissolve gelatin in boiling water. Gradually stir into pie filling. Add juice and 1/2 cup cold water. Pour into 8 by 8 by 2 inch dish. Chill until firm. Blend sour cream and sugar with mixer. Spread over gelatin. Garnish top with a few fresh blueberries.

NEW YORK

"Peek A Blue Berry Farm"
Peek A Blue
See the Blue
Berries
Growing
Up sweetly for you
To pick and eat
Making you healthy and wise
Peek A Blue
Berry and feel free
Nature invites you to
Seek the wonderful life
Of simplicity
God grows berries for us
Our bounty is full
As the moon
Lighting our path
Today
Submitted by Paula Timpson

" I have wonderful memories of enjoying this cake with coffee at my Aunt Lena's kitchen table. My younger brother remembers receiving this cake many times from my Aunt while he was in college and graduate school. She would carefully wrap it in a shoebox and he says that it was perfectly fresh when it arrived. Enjoy!"

Blueberry Sally Lunn

Submitted by: Jean Cole
From: Groton, NY
Bake Time: 50 minutes
Bake Temperature: 350 degrees

Ingredients:

1/2 Cup shortening
1/2 Cup sugar
2 eggs
1 Cup milk

1 3/4 Cup flour
3 Tsp. baking powder
1/2 Tsp. salt

1/4 Cup brown sugar
1/2 Tsp. Cinnamon
2/3 Cup blueberries

Instructions:

Cream together shortening and sugar. Beat eggs; add. Sift together flour, baking powder and salt. Add alternately with milk to creamed mixture. Carefully fold in blueberries. Pour into greased 8 by 8 by 2 inch pan. Mix brown sugar and cinnamon. Sprinkle over batter. Bake at 350 for 50 minutes. Serve hot.

Blue Ambrosia

Submitted by: Joan Compton
From: Stillwater, NY

Ingredients:

1 Cup blueberries
2 Tbls. sugar
1 Cup small green seedless grapes, halved

1 Cup sour cream
2 Cups diced marshmallows

1 Cup chunk pineapple
1/2 Cup shredded coconut

Instructions:

Drain pineapple, then combine with the other ingredients in a bowl. Chill at least 1 hour. Makes 6 servings. For a variation: omit sour cream and marshmallows and add 1 cup white wine. Add coconut just before serving.

Blueberry Pizza

Submitted by: Mrs. Anne Keddy
From: Bath, NY
Bake Time: 15 minutes
Bake Temperature: 400 degrees

Ingredients:

1 1/2 Cups flour
1 8 Oz. Pkg. cream cheese
(1) 8 oz. pkg. whipped topping
3/4 Cup confectioners' sugar
(1) 3 Oz. Pkg. black cherry gelatin (any dark color)

1/2 Cup chopped nuts
1/2 Cup sugar
1/4 Cup sugar
4 Cups blueberries

1 Cup butter
1 Cup water
4 Tbls. cornstarch

Instructions:

Mix together flour, butter, 1/4 cup sugar and nuts (optional) to form dough. Spread in pizza pan or any shape pan. Bake at 400 for 15 minutes. This will look uncooked. Mix cream cheese and confectioners' sugar. Fold in whipped topping and spread over cooled crust. Combine gelatin, 1/2 cup sugar, and 1/2 cup water. Dissolve cornstarch in remaining water; stir in gelatin mixture. Cook over low heat until thickened. Stir in berries to coat all. Cool. (Watch carefully as this can harden quickly.) Spread on top of filling. Chill.

Pumpkin Cake Roll

Submitted by: Anonymous
From: Cattaraugus, NY
Bake Time: 15 minutes
Bake Temperature: 375 degrees

Ingredients:

3 eggs
(2) 3 Oz. pkg. cream cheese
1 Tsp. baking powder
4 Tbls. butter
1 Tsp. lemon juice
1 Cup confectioners' sugar

2 Tsp. cinnamon
3/4 Cup flour
1/2 Tsp. nutmeg
1/2 Tsp. vanilla
1 Cup chopped nuts (optional)

1 Tsp. ginger
2/3 Cup pumpkin
1/2 Tsp. salt
2 Cups fresh blueberries
1 Cup sugar

Instructions:

Beat eggs for 5 minutes. Gradually add sugar. Stir in pumpkin and lemon juice. Combine flour, baking powder, cinnamon, ginger, nutmeg, and salt. Fold dry ingredients into pumpkin mixture. Spread into a wax paper lined 15 by 10 by 1-inch pan. Top with nuts. Bake at 375 degrees for 15 minutes. Turn out on a towel sprinkled with confectioners' sugar. Roll up cake in a towel. Cool. Unroll. Mix 1 cup confectioners' sugar, cream cheese, butter vanilla and blueberries together. Spread over unrolled cake. Reroll and chill for 1 hour or more.

Strawberry Blueberry Freezer Jam

Submitted by: Christine Di Pirro
From: Orchard Park, NY

Ingredients:
1 1/2 Cup crushed strawberries   1 box Sure-Jell Fruit Pectin   1 Cup crushed blueberries
3/4 Cup water   4 1/2 Cups sugar

Instructions:
Mix strawberries, blueberries, and sugar thoroughly. Set aside for 10 minutes, stirring occasionally. Stir in pectin and water in saucepan. Bring to a boil, stirring constantly. Boil 1 minute. Remove from heat. Stir pectin mixture into fruit mixture. Stir constantly until sugar is completely dissolved and no longer grainy, about 3 minutes. Pour into plastic containers to within 1/2 inch of tops; cover. Let stand at room temperature 24 hours. Jam is now ready to use. Store in refrigerator up to 3 weeks or freeze extra containers up to 1 year. Thaw in refrigerator. Makes 5 one cup containers.

Blueberry Flummery

Submitted by: Laura Phillips
From: New York, NY

Ingredients:
2 Cups blueberries   1 Tbls. cornstarch   1/4 Tsp. salt
1/4 Cup water   1/2 Cup sugar   2 Tbls. water
3 Tbls. lemon juice

Instructions:
Combine blueberries and 1/4 cup water in sauce pan and cook over low heat until berries have become soft. Combine cornstarch and remaining 2 tablespoons of water. Add this, the sugar, lemon juice and salt to the blueberries and cook, stirring constantly, until thick and firm. Pour into a serving dish. Chill. Serve with heavy cream. Creates 4 servings.

Blue Wally Pancakes

Submitted by: Ildra Morse
From: Albany, NY

Ingredients:

1 1/2 Cups flour
1 1/2 Cups whole wheat flour
2 1/2 Tsp. baking powder
1 Cup chopped walnuts

1 Tsp. salt
3 large eggs
1 Cup fresh blueberries
1 Tbls. honey

1 Tsp. butter
6 Tbls. butter
3 Cups milk

Instructions:

Combine flours, the baking powder and salt in a large bowl. Mix in blueberries and walnuts; set aside. In another large bowl, lightly beat the eggs. Add the milk and stir to combine Melt 6 tablespoons of butter with the honey in a small saucepan over low heat; stir into the egg mixture. Add the flour mixture to the egg mixture and wisk together until the batter is just smooth. Rest at room temperature for 30 minutes for the batter to aerate. Melt the remaining 1-teaspoon of butter in a medium sized, nonstick skillet over medium heat until it foams. Tilt the skillet; making sure the batter coats the entire surface. Ladle 1/4 cup of the batter into the skillet and cook for about 30 seconds, until small bubbles form on top of the pancake. Turn it over and cook for about 30 seconds, or until golden brown. Repeat with the remaining batter, adding more butter to the pan to prevent sticking. Serve the pancakes as you make them, with syrup; or keep warm, loosely covered with foil, in a low 200-degree oven.

Double Blueberry Pie

Submitted by: Greg Wilson
From: Woodhull, NY

Ingredients:

Whipped topping
4 Cups fresh blueberries

1/4 Tsp. fresh ground cinnamon
1 10 Oz. jar blueberry all-fruit spread

1 baked 9 inch pie shell

Instructions:

In a small saucepan or microwavable bowl, combine fruit spread and cinnamon. Warm over low heat for 2 -3 minutes, stirring constantly, or microwave on high about 1 minute, just until liquefied. In a large bowl, combine berries and fruit spread mixture; spoon into the baked pie shell. Cover loosely with waxed paper and chill until set, 1 to 2 hours. Serve with whipped cream or vanilla ice cream. Creates 6 servings.

Blueberry Frost

Submitted by: Lucille Schuchardt-DeSerio
From: Bath, NY

Ingredients:

1 1/2 Cup blueberries
1 Cup vanilla yogurt
1/2 Tsp. nutmeg
1/4 Cup honey
1/2 Tsp. cinnamon
1 Cup frozen unsweetened sliced peaches
1 Cup milk
Cinnamon sticks

Instructions:

Combine blueberries, peaches and milk in a blender; cover and process on high. Add yogurt, honey, cinnamon and nutmeg; blend well. Pour into glasses. Garnish with cinnamon sticks. Serve immediately. Creates 4 one-cup servings.

Aunt Lena's Blueberry Coffee Cake

Submitted by: Elizabeth Simms
From: Bath, New York
Bake Time: 35 minutes
Bake Temperature: 375 degrees

Ingredients:

3/4 C sugar
2 tsp. Baking powder
2 C flour
3 Tbsp. Flour
1 egg
1/2 Tbsp. Cinnamon
1/2 C sugar
Streusel Mix - 1/4 C Oleo or butter
1/4 Cup Oleo or butter
1/2 tsp. Salt
1/2 C milk

Instructions:

Mix and beat all cake ingredients, then blend in 2 Cups of blueberries. Spread into a greased 9 x 13 pan. Streusels mix - soften butter, combine Streusel ingredients (1/4 c. butter, 1/2 c. sugar, 3 Tps. flour and 1/2 Tpsp. cinnamin) with fork, sprinkle on top before baking. Bake 35 minute or until golden brown on top.

 North Carolina

"There are a couple of blueberry farms in our town. They usually have a blueberry festival around the 1st of June with all kinds of games, hot dogs, and a parade. We also have a town of Bath in NC. Blackbeard was supposed to have lived there at one time…"

Individual Blueberry Cheesecake

Submitted by: Betty Hewitt
From: Hickory, NC

Ingredients:
14 vanilla wafers   1/2 Cup lemon juice   1 8 Oz. pkg. cream cheese
1 1/4 Cups fresh blueberries   1 14 Oz. can sweetened condensed milk

Instructions:
Place cupcake liners in 12 (3 inch) muffin paper cups: place 1 vanilla wafer in each cup. Into a small bowl, crumble the remaining 2 wafers; set aside. In a mixing bowl, with an electric mixer, beat cream cheese until smooth. Gradually beat in condensed milk. Add lemon juice, stirring until mixture starts to thicken. Set aside 12 large blueberries; fold the rest of the blueberries into cream cheese mixture. Spoon mixture into cupcake liners. Top each with some of the reserved crumbs and a large blueberry. Cover loosely with waxed paper; refrigerate in the muffin pan until firm (1 to 2 hours) before serving. Creates 12 servings.

Blueberry Shake

Submitted by: Anonymous
From: Durham, NC

Ingredients:

1 Pint blueberries   1/4 Cup lemon juice   2 Cups ice cubes
1 Cup cold water   1 14 Oz. can sweetened condensed milk

Instructions:

In a blender, combine all ingredients except ice; blend well. Gradually add ice, blending until smooth. Garnish as desired. Refrigerate leftovers. Mixture will stay thick and creamy in refrigerator. Makes about 5 cups.

Blueberry Freezer Pie Filling

Submitted by: Kathy Wilson
From: Sylva, NC
Bake Time: 75 minutes
Bake Temperature: 425 degrees

Ingredients:

1 1/3 Cups sugar   2 Tbls. butter   8 Cups blueberries
1/2 Tsp. salt   1/3 Cup quick cooking tapioca   2 Tbls. lemon juice

Instructions:

In a large bowl, mix sugar, tapioca and salt. Add blueberries and lemon juice; mix gently. Line two 9 inch pie plates with heavy aluminum foil, extending it 6 inches beyond rim. Spoon 1/2 of the mixture into each lined plate; dot each with 1 tablespoon of butter. Fold foil loosely over filling. Place in freezer; freeze 4 hours. Remove filling from pans; seal tightly with foil with foil and return packages of filling to freezer. To use; make pastry for double crust. Unwrap frozen filling; place in pastry lined pie plate. Roll out top crust and cut 2 inch circle from center. Place over filling; seal well. Bake at 425 for 1 hour and 15 minutes. Creates 2 fillings.

Blueberry Crunch Loaf

Submitted by: Judy Dorton
From: Monroe, NC
Bake Time: 60 minutes
Bake Temperature: 350 degrees

Ingredients:

1/4 Cup butter
1 3 Oz. pkg. cream cheese
1 Cup sugar
2 eggs
2 Cups flour

2 Tsp. baking powder
1/2 Tsp. soda
1/4 Tsp. salt
1 Tsp. flour
1/3 Cup milk

3 Tbls. brown sugar
2 Tbls. regular oats
1 Cup blueberries
1 Tsp. butter
2 Tbls. chopped pecans

Instructions:

Beat 1/4 cup butter and cream cheese at medium speed with electric mixer 2 minutes or until creamy. Gradually add 1 cup sugar. Beat at medium speed for 5 minutes. Add eggs, one at a time, beat after each addition. Combine the 2 cups flour, baking powder, soda , and salt. Add to butter mixture alternately with milk, beginning and ending with the flour mixture. Gently fold in berries, spoon batter into greased and floured 9 by 5 by 3 loaf pan. Combine brown sugar, oats, pecans, A.P. flour, and 1 teaspoon melted butter. Stir until mixed. Sprinkle over batter. Bake at 350 for 1 hour or until toothpick comes out clean. Cool in pan on wire rack for 15 minutes. Remove from loaf pan and cool completely.

Blueberry Snack Bars

Submitted by: Steve Shafer
From: Goldsboro, NC
Bake Time: 55 minutes
Bake Temperature: 325 degrees

Ingredients:

3 3/4 Cups flour
1 1/2 Cups butter
3/4 Cup confectioners' sugar
1 Tbls. vanilla

3 Cups blueberries
3 Tbls. lemon juice
1 1/2 Cups coconut
3 Cups sugar

1 1/2 Tsp. cinnamon
1 1/2 Tsp. baking powder
6 eggs
3/4 Tsp. salt

Instructions:

Preheat oven to 325 degrees. Mix one cup flour, softened butter, and powdered sugar until smooth. Spread over bottom of an 8 inch square pan. Bake for 20 to 25 minutes. Mix together eggs, sugar, blueberries, coconut, remaining 1/4 cup flour, lemon juice, vanilla, baking powder, cinnamon, and salt. Spread over the top of the cookie base. Bake for another 30 minutes; cool.

Blueberry Ice Cream

Submitted by: Sandy Chatham
From: Elon College, NC

]Ingredients:

1 Cup water   2 Cups sugar   4 Cups milk
2 Cups blueberries, mashed   (1) 12 Oz. can evaporated milk, chilled
1 large pack vanilla instant pudding

Instructions:

Combine pudding mix and sugar in a large bowl. Add remaining ingredients, stirring well. Pour mixture into freezer can of 1 gallon ice cream freezer. Freeze according to manufacturer's instructions.

Sour Cream Blueberry Cake

Submitted by: Ellen Purifoy
From: New Bern, NC
Bake Time: 45 minutes
Bake Temperature: 350 degrees

Ingredients:

1/2 Cup sweet butter   2 Cups blueberries   3 eggs
1/2 Cup brown sugar   1 1/4 Cup cake flour   1 Cup sour cream
1/2 Tsp. salt   1 Tsp. baking powder   1 Tsp. baking soda
1 Tsp. vanilla   1 Cup sugar

Instructions:

Preheat oven to 350 degrees. Thoroughly grease and flour a 9 by 13 by 2 inch baking pan. Cream butter and white sugar. Add eggs, one at a time, beating well after each addition. Sift dry ingredients and add gradually to the egg mixture alternately with sour cream, ending with flour mixture. Pour half the batter into the pan. Place blueberries on top and sprinkle with brown sugar. Add the rest of the batter and bake for 45 minutes. Cool in the pan 10 minutes. Use a wire rack to finish cooling.

Loads of Blueberries Coffee Cake

Submitted by: Sandy Vartorella
From: Durham, NC
Bake Time: 60 minutes
Bake Temperature: 350 degrees

Ingredients:

4 Tbls. unsalted butter
3 Cups blueberries
2 Cups flour
2/3 Cup sugar

2 1/2 Tsp. baking powder
1/2 Tsp. salt
2 eggs
3/4 Cup milk

1/8 Tsp. cinnamon
2 Tbls. sugar
Pinch of nutmeg
1/4 Tsp. cinnamon

Instructions:

Heat oven to 350 degrees. Grease a 8 or 9 inch square baking pan. In a large, microwave safe bowl, melt butter and set aside to cool. If using fresh blueberries, wash and drain. Spread them on a paper towels to dry, removing any bits of leaf of stalk. In a medium bowl, wisk together flour, baking powder, salt, and 1/8 teaspoon cinnamon. Wisk milk, 2/3 cup sugar, and eggs into the cooled butter. Using a spatula, blend dry ingredients into liquid ingredients. Gently fold in blueberries. Spread batter in prepared baking dish. Blend 2 tablespoons sugar with remaining cinnamon and a pinch of nutmeg. Sprinkle on top of the batter. Bake 45 to 60 minutes or until a toothpick comes out clean. Place dish on a wire rack to cool at least 30 minutes before serving. Cut into squares. Note: best eaten within a day or two of preparation. For longer storage, freeze individual squares and thaw as needed.

Marla's Fresh Blueberry Cream Pie

Submitted by: Aunt Sue Lynch
From: Mebane, NC
Bake Time: 25 minutes
Bake Temperature: 400 degrees

Ingredients:

1 Cup Sour Cream
3/4 Cup sugar
1/4 tsp. Salt
(1) 9" pie pasty unbaked

2 Tbsp. Flour
3 tbs. Flour
2 1/2 Cups fresh Blueberries
Topping - 3 Tbsp. Chopped Pecans or Walnuts

3 Tbsp. Butter softened
1 tsp. Vanilla
1 egg beaten

Instructions:

Combine sour cream, 2 Tbsp. flour, sugar, vailla, salt and beaten egg. Beat for 5 minutes at medium speed or until smooth. Fold in 2 1/2 cups blueberries, pour filling into pastry shell, bake for 25 minutes. Combine 3 Tbls. butter, 3 Tbls. flour and 3 Tbls nuts stirring well. Sprinkle over pie and bake for 10 additional minutes. Chill before serving, makes 4 - 6 servings.

Peek-A-Blue berry farm

NORTH DAKOTA

"The blueberry has a long history in healing. American Indians, who called the fruit star berries for their five pointed blossom, made medicine from the plants' leaves and used blueberry juice to treat coughs…"

Baked Blueberry Rice

Submitted by:	Eileen Will
From:	Elgin, ND
Bake Time:	60 minutes
Bake Temperature:	350 degrees

Ingredients:

1 Cup rice
1 Tsp. vanilla
3/4 Cup sugar
(1) 3.4 Oz. pkg. vanilla instant puddings

3 eggs
1 1/4 Cups milk
Pinch of cinnamon

2 Cups water
1 1/2 Cups blueberries

Instructions:

Boil rice in water, until rice is soft. Mix eggs, milk, vanilla, and sugar. Add the pudding mix and blueberries. Add rice, after draining all the water, to the egg mixture. Pour into a 8 inch square glass pan. Sprinkle with cinnamon and sugar. Bake at 350 for 1 hour. Use this with any meat dish in place of potatoes.

Banana Blueberry Bread

Submitted by: Tina Haugen
From: West Fargo, ND
Bake Time: 60 minutes
Bake Temperature: 350 degrees

Ingredients:

3 Cups flour
1 1/3 Cups sugar
4 Tsp. baking powder

1 Tsp. salt
1 1/2 Cups quick cooking oats
2/3 Cup oil

4 eggs
2 Cups blueberries
2 Cup mashed bananas

Instructions:

Combine all the dry ingredients; stir in oats. Add oil, eggs, bananas and berries, stirring until just mixed. Pour batter into 2 small loaf pans and bake at 350 for 1 hour. Leave pans for 10 minutes after removing form oven. Refrigerate cooled bread before slicing.

Four Fruit Pie

Submitted by: Rebecca Raddohl
From: Rocklake, ND
Bake Time: 60 minutes
Bake Temperature: 400 degrees

Ingredients:

1 Cup rhubarb, chopped
1 Cup blueberries
Butter
1 Cup peeled apple, chopped

3/4 Cup sugar
Pastry for a double crust pie
1 Tsp. lemon juice

1/4 Cup flour
1 Cup raspberries
Sugar

Instructions:

Combine fruits with lemon juice. Stir in sugar and flour. Pour into crust. Dot with butter. Place top crust. Sprinkle with sugar. Bake at 400 for 50 to 60 minutes.

Blueberry Salad

Submitted by: Helen Odegaard
From: Kathryn, ND

Ingredients:
(2) 3 Oz. pkg. grape Jell-O
Small marshmallows

(1) 8 Oz. carton Cool Whip
2 Cups blueberry pie filling

2 Cups boiling water

Instructions:
Mix Jell-O and boiling water. Let cool, then beat in the rest of the ingredients. Place in marshmallows in the mixture. Chill.

Peek-A-Blue berry farm

OHIO

"The secret to successful blueberry freezing is to use berries that are unwashed and completely dry before popping them into the freezer. Completely cover the berry containers with plastic wrap, store in an airtight plastic bag, or arrange dry berries in a single layer on a cookie sheet. When frozen, transfer berries to plastic bags or freezer containers."

Blueberry Swiss Muesix

Submitted by: Marianne Oldja
From: Cleveland, OH

Ingredients:
2 small cartons blueberry yogurt   1/2 Cup nuts, pecans or walnuts   1 Cup fresh blueberries
2 apples, peeled and grated   1 Cup quick oats, uncooked

Instructions:
Mix all ingredients together. Eat right away. Refrigerate remaining cereal.

Blueberry Apple Conserve

Submitted by:　　Garnet Spetnagel
From:　　　　　　Chillicothe, OH

Ingredients:

2 Pints blueberries
1 Cup slivered blanched almonds

1 Tbls. lemon peel
6 Cups sugar

5 Cups chopped apples
1/3 Cup lemon juice

Instructions:

Wash blueberries well in cold water; drain. In a large saucepan, combine with apples, sugar, lemon peel and juice. Cook, stirring constantly, over high heat until sugar is dissolved. Reduce heat; simmer, uncovered and stirring occasionally, 1 1/2 hours or until mixture is very thick. Remove from heat. Add almonds. Meanwhile, sterilize 4 pint jars; leave in hot water until ready to fill. Immediately ladle conserve into hot, sterilized jars, filling to 1/2 inch of the top. Cap at once as manufacturer directs.

Berry Lemonade Slush

Submitted by:　　Maggie Andorka
From:　　　　　　Sheffuld Village, OH

Ingredients:

3 Cups ice cubes
Lemonade Flavor Drink Mix (Country Time)

1 Cup water

1 Cup blueberries

Instructions:

Measure drink mix into cap, just to 1 quart line. Place drink mix, water, ice and fruit in blender container; cover. Blend on high speed about 10 seconds. Turn off blender. Mix with spoon; cover. Blend about 5 seconds until smooth. Serve immediately. Store leftover slush in freezer. Creates 4 servings.

Blueberry Muffins

Submitted by: Jane Kadow
From: Garrettsville, OH
Bake Time: 20 minutes
Bake Temperature: 400 degrees

Ingredients:

1 Cup blueberries
1/4 Cup hazelnuts, chopped
1 Cup sugar
(1) 8 Oz. pkg. cream cheese
2 eggs

1 Tsp. vanilla
1/2 Cup chopped cranberries
1/4 Tsp. nutmeg
1 Tsp. baking soda
1/2 Tsp. salt

1/4 Cup flaked coconut
1 Cup flour
2 Tbls. brown sugar
1/4 Tsp. cinnamon

Instructions:

In a bowl, combine blueberries, cranberries and 1/4 cup white sugar; set aside. In a large mixing bowl, beat cream cheese and remaining sugar until smooth. Add eggs, one at a time, beating after each addition. Beat in vanilla. Combine the flour, baking soda, salt, nutmeg; add to the creamed mixture. Fold in the berry mixture. Fill greased muffin cups 2/3 of the way full. Combine hazelnuts, coconut, brown sugar, and cinnamon. Sprinkle over the batter. Bake at 400 for 18 to 20 minutes or until a toothpick comes out clean. Cool 5 minutes before removing to a wire rack. Creates 1 dozen muffins.

Slow Poke Blueberry Cobbler

Submitted by: Betty Mekus
From: Defiance, OH
Bake Time: 30 minutes
Bake Temperature: 425 degrees

Ingredients:

1/2 Cups sugar
1 Tbls. cornstarch
4 Cups blueberries

1/4 Cup sour cream
2 Tbls. water
1 Cup baking mix

1 Tbls. sugar
1/4 Cup milk

Instructions:

Preheat oven to 425. Grease a 1 1/2 quart casserole dish. Mix 1/2 cup sugar and cornstarch in a 2 quart pan; stir in berries and water. Heat to boiling point, stirring constantly. Boil and stir for 1 minute. Pour into casserole dish. Mix remaining ingredients until soft dough forms. Drop by tablespoonful onto hot fruit mixture. Bake for about 25 to 30 minutes or until golden brown.

Blueberry Gelatin Cake

Submitted by: Elma Austin
From: Twinsburg, OH
Bake Time: 40 minutes
Bake Temperature: 350 degrees

Ingredients:

1 can blueberries   1 box lemon supreme cake mix   Miniature marshmallows
1 6 Oz. pkg. black raspberry Jell-O

Instructions:

Grease the bottom only of a 9 by 13 inch pan. Mix blueberries and juice with dry gelatin. Pour into greased pan. Spread an even layer of marshmallows over all. Mix the lemon supreme cake mix according to package instructions. Spoon batter evenly over marshmallows. Bake at 350 for 35 to 40 minutes. Cool on a rack in the pan for 1 hour. Place in the refrigerator for 3 hours. Gelatin and fruit will form a gel on the bottom while the cake layer and marshmallows will rise during baking. Refrigerate any leftovers.

Blueberry Muffins

Submitted by: Amelia Glavan
From: Lorain, OH
Bake Time: 20 minutes
Bake Temperature: 375 degrees

Ingredients:

1 1/2 Cups corn flakes, crushed   1 Tsp. cinnamon   1 Cup milk
2 Tsp. baking powder   1/2 Cup honey   1 Tsp. baking soda
1 egg, beaten   1 1/4 Cups flour   1 1/4 Cups melted butter
1 Cup blueberries   1/2 Cup white raisins

Instructions:

Heat oven to 375. In a large bowl, soak corn flakes in milk. Stir in honey, egg and melted butter. Add cinnamon, baking powder and baking soda to flour and combine with the above. Add the blueberries and, as an optional feature, add 1/2 cup white raisins. Fill baking cups 3/4 cup full and bake for 20 minutes.

Blueberry Snow Ice Cream

Submitted by: Donna Ransom
From: Sandusky, OH

Ingredients:

2 eggs
1 Cup undiluted evaporated milk

1 Tsp. vanilla
1 Cup blueberries

Snow, new fallen is best
3/4 Cup sugar

Instructions:

Beat eggs; add milk, sugar and vanilla. Fold in snow until mix has the consistency of ice cream. Fold in berries. You might have to place it in the freezer for a bit to get the consistency right.

Blueberry Applesauce Bread

Submitted by: Faye Cromer
From: Troy, OH
Bake Time: 60 minutes
Bake Temperature: 350 degrees

Ingredients:

3 Cups flour, sifted
1 Cup applesauce
1 Tsp. salt
1/2 Cup chopped apricots

2 eggs, beaten
1 Tbls. baking powder
2 Cups fresh blueberries
1/2 Tsp. ground mace

1 Cup sugar
1/4 Cup melted butter
1/2 Tsp. baking soda

Instructions:

Mix flour, sugar, salt, baking soda and mace together in a bowl. In a separate bowl, mix the eggs, applesauce, and butter. Add to dry ingredients, mixing until well blended. Fold in fruit. Grease a 13 by 4 by 3 inch loaf pan and pour in the batter. Bake at 350 for 50 to 60 minutes. Cool for at least 5 minutes before unmolding with a knife. Cut into slices and serve.

Blueberry Topped Sponge Flan

Submitted by: Elma Austin
From: Twinsburg, OH
Bake Time: 25 minutes
Bake Temperature: 350 degrees

Ingredients:

1 Cup sifted flour
1 Tsp. baking powder
1/4 Tsp. salt
1/3 Cup milk
1/2 Cup strawberry Jell-O

1 Tbls. water
1 Tsp. vanilla
3 eggs
1 Cup sugar

7-8 large strawberries
2 Cups fresh blueberries
pastry Cream
2 Tbls. butter

Instructions:

Sift flour, baking powder and salt onto wax paper. Preheat oven to 350. Heat milk with butter just to scalding; cool slightly. Beat eggs, sugar, and vanilla in a medium sized bowl with electric mixer until creamy. Add flour mixture alternately with milk, beating after each addition. Pour into greased and floured 10 inch sponge flan pan. Bake at 350 for 25 minutes or until top springs back when lightly pressed with fingers. Cool in pan on wire rack for 10 minutes. Loosen around edges with a knife; turn out on wire rack; cool. Heat strawberry jelly and water in small saucepan until melted and bubbly. Brush over interior of shell and sides of cake. Allow to set for 5 minutes. Fill center of sponge flan with pastry cream/ pudding. Arrange 8 strawberries in center. Arrange blueberries around top. Glaze berries with additional melted strawberry jelly if desired. Serves 12.

OKLAHOMA

"My daughter made this recipe for her dad when he was ill. He loved them. They are very moist and are a Weight Watchers' recipe. I am a 77-year-old retired school secretary with 2 schoolteacher kids and one that is a Wal-Mart bakery manager. I collect jokes, rocks, small vases, and snowmen…"

Fresh Blueberry Crunch

Submitted by: Sherrill Mercer
From: Grove, OK
Bake Time: 45 minutes
Bake Temperature: 350 degrees

Ingredients:

4 Cups fresh blueberries
1/2 Cup butter

3/4 Cup regular oats
3/4 Cup flour

1 Cup brown sugar

Instructions:

Place berries in a 2 quart baking dish, spreading evenly. Combine other ingredients and sprinkle over berries. Bake at 350 for 45 minutes.

Blueberry Filled Cookies

Submitted by: Tim Flinchum
From: Wagoner, OK
Bake Time: 10 minutes
Bake Temperature: 400 degrees

Ingredients:

1 Cup butter
1/2 Tsp. salt
1/8 Tsp. cinnamon
1/2 Cup milk with 1/2 Tbls. vinegar beat in

1 Tsp. Soda
3 1/2 Cup flour
2 Cups fresh blueberries mashed

2 Cups brown sugar
2 eggs
1 Tsp. vanilla

Instructions:

Mix butter, sugar and eggs. Sift flour, salt, soda and cinnamon. Add vanilla to the batter, then add milk and flour to batter alternately. Drop on ungreased cookie sheet. Add 1/2 teaspoon of mashed blueberries on dough balls and 1/2 teaspoon batter on top. Bake at 400 for 10 minutes.

Diabetic Blueberry Salad

Submitted by: Jean Floyd
From: Tecumseh, OK

Ingredients:

1 8 Oz. pkg. cream cheese
1 can crushed pineapple, undrained
25 pkg. Equal of 1/2 pkg. Equal Recipe

2 cans blueberries, drained
2 large packages sugar free cherry Jell-o

1 16 Oz. can sour cream

Instructions:

Make Jell-O as directed, except using only half of the cold water. Add pineapple and blueberries. Stir and chill. In another bowl, mix cream cheese, sour cream and Equal. Chill. When Jell-O sets, top with cheese mixture. Sprinkle with pecans.

Blueberry Bread

Submitted by: Barbara Blake
From: Tahlequah, OK
Bake Time: 45 minutes
Bake Temperature: 350 degrees

Ingredients:

3 Cups flour
2 Tsp. baking powder
1 Cup crushed pineapple, drained
1/2 Cup flaked coconut
1/2 Tsp. salt

1 1/3 Cups sugar
2/3 Cup shortening
1 Tsp. baking soda
1 1/2 Tsp. lemon juice

1/2 Cup milk
4 eggs
1 Cup chopped nuts
2 Cups fresh blueberries

Instructions:

In a medium bowl, sift flour with baking soda and salt. Set aside. In a large bowl, cream shortening with electric mixer until light and fluffy. Gradually beat in sugar. Stir in eggs, milk, lemon juice and pineapple. Add dry ingredients and mix well. Fold in blueberries, nuts, and coconut. Pour dough into 6 greased and floured 6 by 3 1/4 by 2 1/4 inch pans. Bake for 40 to 45 minutes. Unmold and cool on racks. If desired, drizzle with thin powdered sugar frosting when loaves are completely cool. Wrap in foil and freeze individual loaves or store in airtight container. Frosting - Sift one cup of powered sugar into small bowl. Add 1 tsp. milk and 1 tsp. lemon juice. Stir until smooth, adding milk by 1/2 teaspoons until frosting is desired consistency. Drizzle over cooled loaves.

Peek-A-Blue berry farm

OREGON

"Here's a blueberry recipe from Oregon where we grow quite a lot of blueberries. It's one of my daughter's favorite recipes. Whenever her grandmother comes to visit us, she usually makes it and brings it with her. In fact it has become a family joke that my mom can't come to visit unless she brings the blueberry dish…"

"I worked at a blueberry farm for 10 years as a pruner and scale weigher for the pickers they hired in the summer. They sold their farm to Smuckers Jelly Company two years ago…"

Blueberry Cobbler

Submitted by: B. Johnson
From: Eugene, OR
Bake Time: 40 minutes
Bake Temperature: 375 degrees

Ingredients:

1/2 Cup brown sugar
3 Cups fresh blueberries
1/2 Tsp. Cinnamon

1 Cup flour
1 Cup sugar

1 egg, lightly beaten
1 stick butter

Instructions:

Preheat oven to 375 degrees. Sprinkle all of brown sugar on bottom of 10 inch Pyrex baking pan. Arrange fruit evenly in pan over brown sugar. In a separate bowl, stir together flour, sugar and cinnamon. Add egg. Mix until mixture is crumbly. Sprinkle over fruit. Pour melted butter over top. Bake for 30 to 40 minutes or until bubbly.

Blueberry Doughnuts

Submitted by: Georgia Lovegren
From: Clatskanie, OR

Ingredients:

1 Cup fresh blueberries
2 Cups flour
2 1/2 Tsp. baking powder

2 eggs
1/2 Cup sugar
1/4 Tsp. salt

3 Tbls. melted butter
1/3 Cup milk
1 Tsp. grated lemon rind

Instructions:

Wash fresh blueberries and drain. Sprinkle lightly with flour. Sift together flour, baking powder, sugar and salt. Beat eggs and milk together and stir into flour mixture. Stir in butter and lemon rind. Fold in fresh blueberries. Knead dough lightly and roll out about 1/4 inch thick on well-floured board. Cut with a doughnut cutter and fry in deep hot fat 360 degrees. When doughnuts come to the top (about 5 minutes), turn and brown on other side. Drain on paper towel. Serve plain or dusted with powdered sugar. Creates about 2 dozen doughnuts.

Blueberry-Carrot Picnic Cake

Submitted by: Georgia Lovegren
From: Clatskanie, OR
Bake Time: 60 minutes
Bake Temperature: 350 degrees

Ingredients:

1 Cup flour
1 Cup whole wheat flour
2 Tsp. baking powder
2 Tsp. cinnamon
1/2 Cup brown sugar

1 Tsp. baking soda
1 Cup sugar
1 Tsp. salt
1 Cup corn oil

1 Cup pecans, chopped
2 Cups blueberries
4 medium carrots
4 eggs

Instructions:

Preheat oven to 350. Peel and coarsely grate the carrots. Butter and flour 10 cup bundt pan, preferable nonstick. Mix flours, baking powder, soda, cinnamon and salt together in a medium bowl. Combine both sugars in a large bowl. Add eggs one at a time, whisking until smooth. Whisk in oil. Stir in carrots, blueberries and pecans. Add dry ingredients and fold until just blended.; do not over mix. Batter will be thick. Spoon batter into prepared pan. Bake until cake begins to pull away from the sides of the pan and tester inserted comes out clean, about 1 hour. Cool cake in pan on rack for 20 minutes. Invert onto plate. Cool completely.

Blueberry Syrup

Submitted by: Dorathy Foster
From: Portland, OR

Ingredients:

2 Cups blueberries
1 Cup sugar

1/2 Cup light corn syrup
1/4 Tsp. nutmeg

1 Tsp. cinnamon

Instructions:

Mix all ingredients together. Boil 3 minutes or longer, until berries are tender. Will keep for weeks in the fridge. Use on waffles, pancakes and ice cream.

Blueberry Pound Cake

Submitted by: Virginia Ball
From: Cottage Grove, OR
Bake Time: 60 minutes
Bake Temperature: 325 degrees

Ingredients:

1 Cup butter
2 Cups sugar
with 1/4 cup flour

1 Tsp. vanilla
2 3/4 Cups flour
4 eggs

1/2 Tsp. salt
2 Cups blueberries coated
1 Tsp. baking powder

Instructions:

Cream butter and sugar. Add 1 egg at a time, beat well. Add vanilla. Combine dry ingredients, sift together slowly. Add to the cream mixture. Stir in blueberries. Put the 1/4 cup flour in with the dry mix. Grease a 10 inch tube pan with 2 tablespoons of butter, sprinkle with 1/4 cup sugar. Bake at 325 degrees about an hour. Try sifting brown sugar and using it instead of white- giving it a different flavor.

Blueberry Johnny Cake

Submitted by: Niki Baker
From: Sweet Home, OR
Bake Time: 30 minutes
Bake Temperature: 350 degrees

Ingredients:

1/2 Cup butter
1 1/3 Cups yellow cornmeal
1/2 cup sugar plus 2 Tbls. sugar
2 eggs

1/2 Tsp. baking soda
1 Tsp. baking powder
1 Cup flour

1 Cup buttermilk
1 Cup blueberries
1/2 Tsp. salt

Instructions:

Preheat oven to 350 degrees. Coat skillet with cooking spray; set aside. Cut butter in coarse chunks; place in small bowl. Microwave for 40 seconds, or until butter has melted. In a mixing bowl, Wisk cornmeal, flour, sugar, baking powder, baking soda and salt. Add buttermilk and eggs to the melted butter. Wisk to blend. Toss 1/2 cup blueberries into dry mixture. Using a large spoon, stir in liquid ingredients until just blended. Spread batter in skillet and scatter the remaining blueberries on top. Bake for 30 minutes, or until a toothpick comes out clean.

Blueberry Charlotte Russe

Submitted by: Margaret Teufel
From: Hillsboro, OR

Ingredients:

1 1/2 Cups milk
1/2 Cup cold water
(1) 9 Oz. pkg. frozen whipped topping
2 pkg. Lady fingers, split (3 ounces each)

2 envelopes unflavored gelatin
3 Cups fresh blueberries

1 Pound (56) caramels

Instructions:

Split ladyfingers to line the bottom and sides of an ungreased 9 inch spring form pan. Soak gelatin in cold water. Combine caramels and milk and stir over low heat until smooth and melted. Add softened gelatin and stir until gelatin is dissolved. Chill until mixture thickens slightly. Fold in topping and blueberries. Pour mixture into lined spring form pan. Chill until sides are firm. Remove sides from pan and cut into wedges to serve.

Blueberry Ice Cream Puffs

Submitted by: Margaret Teufel
From: Hillsboro, OR
Bake Time: 45 minutes
Bake Temperature: 350 degrees

Ingredients:

1 Cup water
1/2 Cup butter
1 Cup complete pancake mix (the kind that only needs water)

1/4 Cup sugar
2 Cups blueberries

1 Quart vanilla ice cream
4 eggs

Instructions:

For cream puffs, heat water and butter to boiling in medium saucepan. Add pancake mix; stir well until batter forms a ball. Remove from heat. Add eggs, one at a time, beating after each addition. Drop 1/4 cup 3 inches apart onto an ungreased cookie sheet. Bake 45 minutes or until golden brown. To make topping, combine 1 cup blueberries and sugar in blender and process for 10 seconds. Stir in remaining blueberries. Slice off top of puffs. Fill with a scoop of ice cream. Spoon blueberry sauces overtop.

Peek-A-Blue berry farm

PENNSYLVANIA

"Blueberry Jam was introduced into the highlands of Scotland by French cooks in the court of James the 5th."

Blueberry Pie

Submitted by: April Burdette
From: Lebonon, PA
Bake Time: 50 minutes
Bake Temperature: 375 degrees

Ingredients:

4 Cups blueberries   1/4 Tsp. almond extract   1 pie crust pastry
1/2 Cup sugar   1/2 Cup flour   2 Tbls. butter
3 Tbls. flour   1/2 Cup brown sugar

Instructions:

In a mixing bowl combine blueberries, white sugar, 3 tablespoons flour and almond extract. Transfer mixture to a pastry lined 9 inch pie plate. Combine 1/2 cup flour and brown sugar. Cut in 2 tablespoons of butter until mixture resembles coarse crumbs. Sprinkle over filling. Cover edge of pie with foil. Bake at 375 degrees for 25 minutes. Remove foil and bake for an additional 20 to 25 minutes. Crust should be golden and fruit tender.

Blueberry Cobbler

Submitted by: Deb Derstine
From: Shade Gap, PA
Bake Time: 60 minutes
Bake Temperature: 350 degrees

Ingredients:

2 1/2 Cups fresh blueberries
1/2 Cup butter
1/2 Cup sugar
1/4 Tsp. salt

1/2 Cup milk
1 Cup flour
1 Tsp. baking powder

1/2 Tsp. vanilla
1 Cup sugar
1/2 Cup water

Instructions:

In a 10 by 6 inch pan, spread the blueberries. In a mixing bowl combine the stick of butter and 1/2 cup sugar. Add the milk, flour, baking powder, salt and vanilla. Spread evenly over berries. Mix together water and 1 cup sugar. Pour over batter. Bake at 350 for 1 hour.

Frozen Blueberry Fluff

Submitted by: Martha Sensenig
From: Ephrata, PA
Bake Time: 8 minutes
Bake Temperature: 350 degrees

Ingredients:

2 Cups crisp graham crackers
5 Tbls. butter
1 Cup heavy cream, whipped

1 1/2 Cups sugar
1 Tbls. vanilla

2 Cups fresh blueberries
2 eggs whites

Instructions:

Mix cracker crumbs and butter. Press in the bottom of butter 9 by 13 inch pan. Bake at 350 for 8 minutes. Beat egg whites and vanilla slightly in large bowl. Gradually beat in sugar and berries. Beat at high for 12 to 15 minutes. Mixture is fluffy and has a large volume. Fold in whipped cream. Spread over crumb crust. Freeze overnight.

Steamed Blueberry Mush

Submitted by: Joni Dalmas
From: New Holland, PA

Ingredients:

4 Cups blueberries
2 Cups sugar
1 Tbls. butter

2 Cups flour
2/3 Tsp. salt
3 1/2 Tsp. baking powder

3/4 Cups milk
1 Tsp. lemon juice

Instructions:

Sift flour, baking powder and salt together. Add butter and work it into the dry ingredients. Add milk and beat until thoroughly mixed. Add sugar and lemon juice to berries. Fold into batter, stirring just enough to blend together. Pour into a buttered mold, cover tightly and steam for 45 minutes. Serve warm with rich milk or cream. Makes 6 servings.

Blueberry Ice Cream

Submitted by: Wendy Rosenberry
From: Chambersburg, PA

Ingredients:

4 Cups blueberries
4 Cups half-and-half cream

2 Tbls. water

2 Cups sugar

Instructions:

In a large saucepan, combine the blueberries, sugar and water. Bring to a boil. Reduce heat; simmer, uncovered, until sugar is dissolved and berries are softened. Strain mixture; discard seeds and skins. Stir in cream. Cover and refrigerate overnight. Fill cylinder of ice cream freezer 2/3 full; freeze according to manufacture's directions. Refrigerate remaining mixture until ready to freeze. Allow to ripen in ice cream freezer or firm up in the freezer for 2 to 4 hours before serving. Creates about 1 3/4 quarts.

Blueberry Crisp Pudding

Submitted by: Marilyn Carroll
From: West Finley, PA
Bake Time: 40 minutes
Bake Temperature: 375 degrees

Ingredients:

4 Cups fresh blueberries
3/4 Cup quick cooking oats
1/3 Cup brown sugar

1/3 Cup sifted flour
1/3 Cup sugar

4 Tbls. butter
2 Tsp. lemon juice

Instructions:

Place berries in a 1 1/2 quart baking dish. Sprinkle with white sugar and lemon juice. Cream butter; gradually add brown sugar. Blend in flour and oats with fork. Spread topping over blueberries. Bake at 375 for 35 to 40 minutes. Serve plain or with whipped cream. Creates 6 servings.

Banana Blueberry Shake

Submitted by: Annette Banet
From: Lansdale, PA

Ingredients:

(1) 8 Oz. can strawberry yogurt
1/2 Cup frozen blueberries

1 Cup milk

1 medium banana

Instructions:

Blend all ingredients in a blender. Serve immediately or store in refrigerator until ready to serve. Creates 1 shake.

Blueberry Corn Muffin Cake

Submitted by: Joyce Andrews
From: Blossburg, PA
Bake Time: 65 minutes
Bake Temperature: 375 degrees

Ingredients:

2 Cups flour
1 Cup yellow cornmeal
1 1/2 Cups buttermilk
1/3 Cup orange marmalade
3/4 Cup unsalted butter

3 eggs
1 Tbls. finely shredded orange peel
3 Tbls. butter
3 Cups blueberries
1 1/2 Cups sugar

4 Tsp. whipping cream
2 Tsp. baking powder
1/4 Tsp. salt
3 Tbls. sugar

Instructions:

Sift together flour, cornmeal, baking powder and salt. Set aside. In a large mixer bowl, beat together the 3/4 cup butter and 1 1/2 cups sugar until fluffy. Add eggs, one at a time, beating 1 minute after each. Stir in orange peel. Add the cornmeal mixture alternately with buttermilk to egg mixture. Beat at low speed until just combined. Fold in blueberries. Turn batter into a greased and floured 10 inch fluted tube pan, spreading evenly. Bake at 375 for 55 to 65 minutes or until cake tests done. Cool 15 minutes in pan. Remove and cool completely on wire rack. In a small heavy saucepan, combine marmalade, 3 tablespoons sugar, 3 tablespoons butter and whipping cream. Bring to boiling, stirring to dissolve sugar. Boil gently, uncovered, until thickened, about 5 minutes. Cool and spoon over cake, allowing glaze to flow down sides.

Peek-A-Blue berry farm

RHODE ISLAND

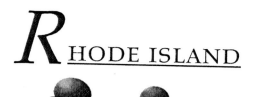

"We had a pick your own blueberry patch a number of years ago. Many people asked for recipes- to use both fresh and from the freezer later on. The number one request was pie-of course!"

Blueberry Pie

Submitted by: Ella Kampper
From: Warwick, RI

Ingredients:

4 Cups blueberries
1 Tbls. lemon juice or vinegar
1/4 Cup water

1 Tbls. butter
3 Tbls. cornstarch
3/4 Cup sugar- part brown sugar

Whipped cream
1 baked 9 inch pie shell

Instructions:

Combine sugar, cornstarch, water, and two cups blueberries in saucepan, cook over medium heat- stirring constantly until mixture comes to a boil and is thickened and clear (will be thick). Remove from heat, stir in butter and lemon juice or vinegar. Cool to almost lukewarm. Place remaining 2 cups of blueberries in pie shell; top with cooked mixture. Spread to cover. Chill. Serve garnished with whipped cream, which is attractive just as a border.

Blueberry Cobbler

Submitted by: Elizabeth Briggs
From: Portsmouth, RI
Bake Time: 25 minutes
Bake Temperature: 425 degrees

Ingredients:

2 Tbls. biscuit mix
1 Cup biscuit mix
1/4 Cup butter
3 Tbls. Boiling water

dash of nutmeg
1 Cup sugar
1/2 Tsp. cinnamon

4 Cups blueberries
1 Cup water
1 Tbls. lemon juice

Instructions:

Heat oven to 425. Mix 2 tablespoons biscuit mix with sugar, cinnamon, nutmeg, 1 cup water, lemon juice and blueberries. Pour into a 1 1/2 quart round baking dish. Stir 1 cup biscuit mix with butter and boiling water until mixture forms a ball. Divide into 8 parts, pat into squares to cover fruit mixture. Bake about 25 minutes, or until dough squares are nicely browned. Serve warm with cream, ice cream, or a whipped topping.

Hot Blueberry Cake with Sauce

Submitted by: Ruth Cook
From: Pascoag, RI
Bake Time: 40 minutes
Bake Temperature: 350 degrees

Ingredients:

1/2 Cup sugar
2 1/2 Cups sifted flour
1 Tsp. nutmeg
1 Tsp. cloves
1 Pint blueberries
1 Tsp. cider vinegar
1/2 Tsp. cinnamon

1 egg
1/2 Cup butter
1 Cup boiling water
1 Tsp. ginger
1 Cup water
2 Tbls. cornstarch

1 Tsp. salt
1 Tbls. Soda
1 Cup dark molasses
1 Cup blueberries
1/2 Cup sugar
1/4 Cup cold water

Instructions:

Cream 1/2 cup sugar and butter together. Stir in molasses and beat in egg. Sift together soda, salt, cloves, nutmeg, ginger, and flour. Beat them into first mixture; then slowly add boiling water. Beat thoroughly. Fold in 1 cup berries. Pour into greased, floured 9 inch pan. Bake at 350 for 40 minutes. Put into saucepan 1 pint blueberries, 1 cup water, 1/2 cup sugar, cider vinegar, and cinnamon. Bring to a full boil. Blend cornstarch with cold water. Add slowly to berry mixture. Cook until clear. Serve over warm over the cake.

SOUTH CAROLINA

"My daughter baked this bread and entered it in the 4-H fair competition in August of 1976. She won first prize for her age division, and then won the grand champion trophy for the entire baking division of the fair competition that year. She picked the blueberries herself, and of course baked the bread herself. She was only 5 years old and this was her first year in 4-H…"

Blueberry Salad

Submitted by: Mary Woiczechowski
From: Aiken, SC

Ingredients:

(1) 8 Oz. pkg. blackberry gelatin
(1) 8 Oz. pkg. raspberry gelatin
(1) 8.5 Oz can crushed pineapple
(1) 15 Oz. can blueberry pie filling

(1) 8 Oz. pkg. cream cheese
2 Cups boiling water
1/2 Tsp. vanilla

1/2 Cup sugar
1 Cup sour cream
Chopped nuts

Instructions:

Dissolve gelatin in 2 cups boiling water. Drain pineapple and berries and add enough water to juice to make 1 cup liquid. Add to gelatin mixture. Stir in berries and pineapple. Pour into 2-quart flat dish and refrigerate until firm. Blend cream cheese, sugar, sour cream, and vanilla. Spread over salad. Sprinkle top with chopped nuts.

Warm Blueberry Sipper

Submitted by: Donna Curry
From: Greenville, SC

Ingredients:
3 Cups water   2 lemon slices 1/4 inch thick   Whipped cream
2 Cups fresh blueberries   1 Tbls. cornstarch   1/4 Cup cold water
1/2 Cup sugar   1 cinnamon stick

Instructions:
In a medium saucepan, combine water, blueberries, sugar, lemon slices and cinnamon stick; bring to a boil. Reduce heat. Simmer 10 minutes or until berries are soft. Remove lemon slices and cinnamon stick. Pour mixture into blender and process until smooth. Return to saucepan. In a small bowl, combine cornstarch and 1/4 cup cold water. Pour into blueberry mixture. Cook over medium heat until mixture boils and thickens, stirring continually for 1 minute. Pour into 8 heatproof glasses. Add a dollop of whipped cream and serve warm.

Blueberry Dessert Bars

Submitted by: Armena Johnson
From: Due West, SC
Bake Time: 20 minutes
Bake Temperature: 450 degrees

Ingredients:
2 Tsp. sugar   (1) 21 Oz. can blueberry pie filling
(1) 20 Oz. pkg. refrigerated sugar cookie dough

Instructions:
Set aside 1/2 of the sugar cookie dough. Pat remainder into 9 by 13 inch pan. Spoon blueberry filling over dough. Scatter small pieces of the remaining sugar cookie dough over the filling. Sprinkle with sugar. Bake at 450 for 20 minutes, or until brown. Cool and cut into square. Makes 24 squares.

Fresh Blueberry Fruit Salad

Submitted by: Lois Allen
From: Charleston, SC

Ingredients:

2 Cups fresh peaches, sliced
2 Cups fresh blueberries
1 Cup pineapple chunks, drained
1 Cup fresh strawberries, sliced
1/2 Cup sugar
1 Cup white seedless grapes
1 Tbls. cornstarch
1 Tbls. lemon juice

Instructions:

Mix peaches, cornstarch, sugar and lemon juice. Cook to a paste, stirring as needed. Chill completely. Prepare remaining fruit and chill. Gently combine the cold peach filling with the blueberries, strawberries, grapes and pineapple chunks. Serve immediately. Makes 4 to 6 servings.

Blueberry Tea Cake

Submitted by: Wanda Le Mieux
From: Anderson, SC
Bake Time: 25 minutes
Bake Temperature: 400 degrees

Ingredients:

1 egg, beaten
2/3 Cup sugar
1 1/2 cups flour, sifted
2 Tsp. baking powder
1/2 Tsp. cinnamon
1/3 Cup milk
3/4 Tsp. salt
1 Tsp. vanilla
3 Tbls. butter
1 Cup fresh blueberries
2 Tbls. sugar

Instructions:

Preheat oven to 400. Grease 1 1/2 quart, shallow baking dish. In a medium bowl, with wooden spoon, beat egg. Gradually beat in 2/3 cup sugar; beat until well combined. Sift together flour, baking powder, cinnamon, and salt. Add to sugar mixture alternately with milk. Beat well after each addition. Add butter and vanilla. Beat well. Fold in blueberries. Pour batter into prepared dish. Sprinkle top with 2 tablespoons sugar. Bake 25 minutes, or until top springs back when touched lightly with fingertip. Serve warm with butter. Makes 8 servings.

Blueberry Gateau

Submitted by:	Dottie Blumberg
From:	Greer, SC
Bake Time:	60 minutes
Bake Temperature:	350 degrees

Ingredients:

1 Cup flour
1 Tsp. Flour
1 Tsp. baking powder
1 Tbls. Confectioners' sugar

1/2 Cup unsalted butter
1 Cup sugar
2 eggs
2 1/2 Cups fresh blueberries

1 Tbls. sugar
1/2 Tsp. lemon juice
1/8 Tsp. salt

Instructions:

Preheat the oven to 350. Using a small brush, lightly coat a 9 inch spring form pan with softened butter or vegetable oil cooking spray. Dust with flour and tap out any excess. Set aside. In a medium-size bowl, sift 1 cup flour, baking powder and salt. Set aside. In a large bowl using a mixer set on medium speed, beat the butter and 1 cup sugar until fluffy and light. Add the eggs, one at a time, and continue to beat until blended. Reduce mixer speed to low and gradually add the flour mixture. Beat just until flour is incorporated. Pour batter into prepared pan and spread evenly. In a medium bowl, toss the berries with the remaining flour, sugar and lemon juice. Spoon berry mixture evenly over batter. Bake on the middle rack of the oven until a toothpick inserted comes out clean, about 1 hour. Cool in the pan on a wire rack. Use a knife to loosen the cake from the pan sides and release cake from the spring form. Transfer to a cake plate, berry side up, and serve warm. Sprinkle with confectioners' sugar.

South Dakota

"According to tribal medicine men, the leaves of blueberry plants were seeped to make a blood purifier…"

Tri- Berry Jam

Submitted by: Mary Hafner
From: Mitchell, SD

Ingredients:

4 Cups blueberries
2 1/2 Cups red raspberries
(2) 1 3/4 Oz. pkg. powdered fruit pectin

1/4 Cup lemon juice
2 1/2 Cups strawberries

11 Cups sugar

Instructions:

Combine the berries and lemon juice in a large kettle; crush fruit slightly. Stir in pectin. Bring to a full rolling boil over high heat, stirring constantly. Stir in sugar; return to a full rolling boil. Boil 1 minute, stirring constantly. Remove from the heat; skim off any foam. Pour into hot jars, leaving 1/4 inch head space. Adjust caps. Process for 15 minutes in a boiling water bath. Creates about 6 pints.

Quick Blueberry Biscuits

Submitted by: Nellie Detmers
From: Canton, SD
Bake Time: 12 minutes
Bake Temperature: 400 degrees

Ingredients:

2 Tbls. sugar
1 Tbls. grated lemon peel

1/2 Cups blueberries
(1) 9.5 Oz. pkg. refrigerated flaky biscuits

Melted butter

Instructions:

With fingers, separate each biscuit into halves horizontally. Combine blueberries, sugar, and lemon peel. Spoon a tablespoon of blueberry mixture in center of 1/2 of each biscuit; top with remaining half. Pinch edges of biscuit together to seal. Repeat with remaining biscuits. Place in greased muffin cups. Brush tops with melted butter. Sprinkle with additional sugar. Bake at 400 for 12 minutes or until golden brown. Creates 10 biscuits.

Best Blueberry Shortcake

Submitted by: Maxine Hohbach
From: Plankinton, SD
Bake Time: 35 minutes
Bake Temperature: 350 degrees

Ingredients:

3 Tbls. butter
1/2 Cup milk
1/4 Cup sugar
1 1/3 Cups flour

3 Tbls. lemon juice
3 Cups blueberries
1 egg, beaten
1/3 Cup butter

2 Tsp. baking powder
3 Tbls. flour
2/3 Cup sugar

Instructions:

Combine butter, blueberries, 2/3 cup sugar, lemon juice and 3 tablespoons flour. Heat to boiling. Pour into an 8 by 8 by 2 inch pan. Mix the remaining ingredients. Drop by spoonfuls over hot blueberry mixture. Bake at 350 for 35 minutes. Serve warm with whipped cream or ice cream.

Blueberry Pie

Submitted by: Luella Morse
From: Hot Springs, SD
Bake Time: 45 minutes
Bake Temperature: 425 degrees

Ingredients:

2 Cups flour
6 Cups blueberries
1/2 Cup flour
1 Cup unsalted butter, very cold

6 to 9 Tbls. ice water
1/2 Tsp. salt
1 Tbls. lemon juice
1 Tbls. butter

1/2 Tsp. cinnamon
2 Tsp. lemon juice
3/4 Cup sugar

Instructions:

In a bowl, mix 2 cups flour and salt; then, using a pastry blender, shred the butter into the flour until mix looks like little peas. Mix in the water and 2 teaspoons lemon juice (adding just enough water for the dough to hold together), and mold the dough into two, flat disks. Refrigerate for half an hour before rolling out. Mix 3/4 cup sugar, 1/2 cup flour and cinnamon in a large bowl. Stir in blueberries, Turn into pastry lined plate. Sprinkle with 1 tablespoon lemon juice. Dot with butter. Cover with top pastry that has slits cut in it, seal and flute. Cover edge with 2 to 3 inch strip of aluminum foil to prevent excessive browning. Remove foil during last 15 minutes of baking. Bake 35 to 45 minutes at 425 degrees. Cool in pie plate on wire rack. Serve warm if desired.

Peek-A-Blue berry farm

TENNESSEE

"This recipe is from the "Presley Family Cookbook." It's by Elvis's uncle and his cook Nancy Rooks. I thought you might be interested to know that in it is the recipe of the last meal Elvis ate. It was a meatball and spaghetti dinner. Also Elvis's favorite sandwich was peanut butter and banana…"

"I started collecting recipes and cookbooks when I was 8 years old, and now I am 61 and still collecting. I currently have 344 cookbooks and boxes and boxes of recipes clipped from paper and magazines, plus I recently inherited all my mother's handwritten recipes from 1899-2002…"

Lemon Blueberry Bliss

Submitted by: Gladys Denning
From: Old Hickory, TN

Ingredients:

1 Cup sugar
3 Cups cooked medium grain rice
1 Tbls. finely grated lemon zest

1/2 Cup butter, softened
2/3 Cup fresh lemon juice
1 Cup fresh blueberries

2 eggs plus 2 eggs yolks
2/3 Cup whipping cream

Instructions:

Combine sugar, eggs, egg yolks, lemon juice and lemon zest in a 2 quart saucepan. Cook over medium-low heat until thick and creamy (8 to 10 minutes), stirring constantly. Remove from heat; stir in butter and rice. Cool. Fold in 1 cup whipping cream. Alternate layers of blueberries and rice pudding in parfait glasses. Garnish with whipped cream and blueberries.

Blueberry Oatmeal Cookies

Submitted by: Neda Hinson
From: Trimble, TN
Bake Time: 10 minutes
Bake Temperature: 375 degrees

Ingredients:
1 Tbls. milk   1 egg   1/4 Cup brown sugar
3/4 Cup quick-cooking oats   1/3 Cup cooking oil
1 Pkg. Duncan Hines Wild Blueberry Muffin Mix

Instructions:
Preheat oven to 375 degrees. Wash blueberries; drain on paper towels. In a medium bowl, combine all but blueberries; mix well. Drop from a teaspoon onto an ungreased cookie sheet. Make a deep depression in the center of each cookie and fill with 7-8 well-drained blueberries. Push dough from sides to cover berries and pat down. Bake for 8 to 10 minutes, until light brown.

Blueberry Pecan Cobbler

Submitted by: Elizabeth Stewart
From: Cookeville, TN
Bake Time: 30 minutes
Bake Temperature: 475 degrees

Ingredients:
4 Pints blueberries   1/2 Tsp. cinnamon   1 Tsp. vanilla
1 1/2 Cups sugar   1/3 Cup water   (1) 15 Oz. pkg. refrigerated
Pie crusts   1/2 Cup flour   2 Tbls. lemon juice
1/2 Cup chopped pecans, toasted

Instructions:
Bring blueberries, sugar, flour, cinnamon, water, lemon juice and vanilla to a boil in a saucepan over medium heat, stirring until sugar melts. Reduce heat to low; cook, stirring occasionally, 10 minutes. Spoon half of the blueberry mixture into a lightly greased 8 inch square pan. Roll 1 piecrust to 1/8 inch thickness on a lightly floured surface; cut into an 8 inch square. Place over blueberry mixture, sprinkle with pecans. Bake at 475 for 10 minutes. Spoon remaining blueberry mixture over baked crust. Roll remaining piecrust to 1/8 inch thickness; cut into 1 inch strips. Arrange in lattice design over blueberry mixture. Bake at 475 for 10 minutes or until golden. Serve with vanilla ice cream. Creates 4 servings.

Blueberry Raspberry Crunch

Submitted by: Zelma Clough
From: Harriman, TN
Bake Time: 30 minutes
Bake Temperature: 375 degrees

Ingredients:

1 21 Oz. can blueberry pie filling   1/2 Cup chopped nuts   1 white cake mix
1/2 Cup butter, melted   1 21 Oz. can raspberry pie filling

Instructions:

Combine pie fillings in a greased 13 by 9 by 2 inch baking dish. In a bowl, combine cake mix, nuts and butter until crumbly; sprinkle over filling. Bake at 375 for 25 to 30 minutes or until filling is bubbly and topping is golden brown. Serve warm. Creates 12 servings.

Blueberry Lemon Pudding Cake

Submitted by: Vicki Ellison
From: Fayetteville, TN
Bake Time: 35 minutes
Bake Temperature: 350 degrees

Ingredients:

1/4 Cup flour   1 Cup low fat buttermilk   2 Tbls. butter
2/3 Cup sugar   1 Tsp. grated lemon rind   2 egg yolks
1/8 Tsp. salt   3/4 Tsp. confectioners' sugar   3 egg whites
1 1/2 Cups blueberries   1/4 Cup fresh lemon juice   1/4 Cup sugar
1/8 Tsp. ground nutmeg

Instructions:

Preheat oven to 350. Combine the flour, 2/3 cup sugar, salt, and nutmeg in a large bowl; add the buttermilk, lemon rind, lemon juice, butter, and egg yolks, stirring with a wisk until mixture is smooth. Beat egg whites with a mixer at high speed until foamy. Add 1/4 cup sugar, one tablespoon at a time, beating until stiff peaks form. Gently stir 1/4 of egg white mixture into the buttermilk mixture; gently fold in remaining egg white mixture. Fold in blueberries. Pour the batter into an 8 inch square baking pan coated with cooking spray. Place in a larger baking pan, add hot water to larger pan to a depth of 1 inch. Bake at 350 for 35 minutes or until cake springs back when touched lightly in center. Sprinkle cake with powdered sugar. Serve warm. Creates five one cup servings.

Blueberry Oat Muffins

Submitted by: Joann Forgason
From: Michie, TN
Bake Time: 20 minutes
Bake Temperature: 375 degrees

Ingredients:

1 Cup flour
1 Cup oats
1/4 Cup vegetable oil
1 Cup blueberries

2 Tsp. baking powder
1/2 Tsp. salt
2 Tbls. brown sugar

1/4 Cup milk
1/2 Cup maple syrup
2 egg whites, beaten

Instructions:

Combine flour, oats, brown sugar, baking powder, and salt. Stir in egg whites, maple syrup, oil, and milk. Fold in blueberries. Fill greased muffin tins and bake at 375 for 20 minutes.

Blueberry Cheesecake

Submitted by: Elizabeth Stewart
From: Cookeville, TN
Bake Time: 70 minutes
Bake Temperature: 300 degrees

Ingredients:

1 1/2 Cups finely ground almonds
1/4 Cup sugar
3 Tbls. butter
1 1/2 Cups blueberries
1/2 Tsp. salt
1 Tbls. flour

1 1/4 Cups sugar
3 Tbls. flour
(3) 8 Oz. pkg. cream cheese
1 Cup whipping cream
2 Tbls. sour cream

4 eggs
(1) 8 Oz. can sour cream
1 Tsp. vanilla
2 Tsp. sugar
1 Tbls. grated lemon rind

Instructions:

Combine almonds, 1/4 cup sugar, butter and 1 tablespoon flour in a small bowl. Press mixture into the bottom and 1 1/2 inches up sides of a lightly greased 9 inch spring form pan; set aside. Beat cream cheese at medium speed with an electric mixer until smooth. Combine 1 1/4 cups sugar, 3 tablespoons flour, and salt. Add to cream cheese, beating until blended. Add eggs, one at a time, beating well after each addition. Add 8 ounce container of sour cream, vanilla, and lemon rind, beating just until blended. Gently stir in blueberries. Pour mixture into prepared pan. Bake at 300 for 1 hour and 10 minutes or until center is firm. Turn off oven. Let cheesecake stand in oven, with oven door partially open, 30 minutes. Cover and chill 8 hours. Release sides of pan. Beat whipping cream at high speed until foamy; gradually add 2 teaspoons sugar, beating until stiff peaks form. Fold in 2 tablespoons sour cream. Spread over cheese cake, and garnish. Creates 12 servings.

TEXAS

"If a recipe calls for a can of blueberries, you may make your own by using 2 ½ cups of fresh blueberries, 1 tablespoon cornstarch, 1 ½ teaspoons lemon juice, 1/8 cup water, cook until thickened and clear…"

"I'm a baker in Texas and I like trying out new recipes too, and I haven't had anything with blueberries since my grandma died. I used to go out in the field and get her a 5 gallon bucket of the juiciest berries I could find. She made an awesome blueberry cobbler…"

Texas Blueberry Syrup

Submitted by: Mae Dobbins
From: Ore City, TX

Ingredients:

1 1/2 Cups blueberries   1 Tbls. cornstarch   1 Cup light corn syrup

Instructions:

Place 1 cup blueberries and corn syrup in a blender. Blend on high for 30 seconds. In a medium saucepan, stir blueberry mixture into cornstarch. Cook and stir over medium heat. Bring to a boil and boil one minute. Remove from heat. Stir in remaining blueberries. Store in the refrigerator.

Mexican Cobbler

Submitted by: Ruth Ann Blake
From: Coleman, TX
Bake Time: 45 minutes
Bake Temperature: 300 degrees

Ingredients:

2 cans blueberry pie filling
1 1/2 Cups water

1 1/2 Cups margarine
2 Cups sugar

8 large flour tortillas

Instructions:

Place filling into tortillas and roll up. Put into baking dish (at least 9 by 13, large would be better). Place the sugar, margarine and water into a saucepan. Cook until boiling. Pour mixture over tortillas. Let soak 4 to 6 hours or overnight. Bake at 300 until bubbly and brown, about 40 to 45 minutes. Serve topped with whipped cream.

Blueberry Chocolate Puffs

Submitted by: Nancy Karr
From: Linden, TX
Bake Time: 15 minutes
Bake Temperature: 425 degrees

Ingredients:

2 Tbls. vegetable shortening
1 egg
(1) 17 oz. box frozen puff pastry sheets

1 Tsp. water
3 3oz. squares semi-sweet chocolate

2/3 Cup blueberry jam

Instructions:

Thaw pastry for 20 minutes, then unfold. Roll each square until 10 inches square in size. Cut each sheet into 20 (2inch) rounds. Place round on ungreased baking sheets. Combine water and egg and brush top of each round. Chill for 10 minutes, then bake at 425 for 12 to 15 minutes. Remove to wire rack and cool completely. Split each puff and spread bottom half with 1/2 teaspoon blueberry jam, then put top half back in place. Melt chocolate; add shortening and stir until smooth. Drizzle top of each puff and let set until chocolate hardens.

Very Berry Shortcakes

Submitted by: Nancy Karr
From: Linden, TX
Bake Time: 12 minutes
Bake Temperature: 425 degrees

Ingredients:

2 Cups strawberries
1 Cup raspberries
1 Cup blueberries
3 Tbls. butter, melted

3 Tbls. sugar
2 1/3 Cups Bisquick baking mix
1/2 Cup milk

3/4 Cup sour cream
2 Tbls. brown sugar
1/3 Cup sugar

Instructions:

Sprinkle fruit with 1/3 cup sugar; let stand 1 hour. Heat oven to 425 degrees. Mix baking mix, milk, 3 tablespoons sugar and the butter until soft dough forms. Smooth dough into ball on cloth covered board lightly floured with baking mix. Knead 8 to 10 times. Roll dough 1/2 inch thick. Cut with floured 3 inch cutter. Bake on ungreased cookie sheet until golden brown, 10 to 12 minutes. Mix sour cream and brown sugar. Split shortcakes; fill and top with fruit. Top with sour cream mixture.

Blueberry Angel Food Cake

Submitted by: Mildred Keller
From: Orange, TX

Ingredients:

2 Cups Blueberries
1 Cup confectioners' sugar
(1) 8 Oz. pkg. cream cheese
1 baked Duncan Hines angel food cake mix

1 Cup sugar
1 Tbsp. cornstarch
2 pkg. Dream Whip

1 Stick butter
1 Cup sugar

Instructions:

Bake the cake mix as directed. Freeze for 1 hour. Slice in half with electric knife. Cream together 1 cup sugar, powdered sugar and cream cheese. In another bowl mix Dream Whip as directed. Do not use Cool Whip. Fold Dream Whip mixture into the cheese mixture. Cook the blueberries over the stove for 15 minutes. DO not add water. When they have fluid in them, thicken with cornstarch, 1 cup sugar and stick of butter. Let blueberries set overnight. Cover bottom layer with Dream Whip mixture. Then add blueberry mixture all around cake. Put on top layer and do the same thing. Let the berries run down sides of the cake about every 2 inches for a pretty effect. Let set in refrigerator for 2 hours before serving cake, will last in the refrigerator for 4 days.

Blueberry Topping

Submitted by: Mary Lynn Raybourn
From: Gary, TX

Ingredients:

1 1/2 Cups blueberries   Confectioners' sugar to taste

Instructions:

Place blueberries in blender with confectioners' sugar. Blend until smooth and fluffy. Serve on waffles, pancakes or ice cream.

Texas Blueberry Cheesecake

Submitted by: Mrs. Troy Parker
From: Lone Star, TX
Bake Time: 10 minutes
Bake Temperature: 450 degrees

Ingredients:

1 Cup flour   1 Cup chopped pecans   1 Tsp. lemon juice
1 Cup sugar   1/4 Tsp. salt   1 8 Oz. pkg. cream cheese
2 Cups blueberries   2 Tbls. cornstarch   1 Stick butter
1 1/2 Cup sugar   2 Cups whipped topping

Instructions:

Blend together flour, salt and butter. Stir in chopped nuts. Press into 9 inch pie pan. Bake at 450 for 10 minutes or until done. Blend cream cheese with 1 1/2 cup sugar. Fold in whipped topping. Pour into crust. Combine blueberries, 1 cup sugar and cornstarch in a saucepan. Cook until thick, add lemon juice. Cool and pour over pie.

Blueberry Bars

Submitted by: Imogene Powers
From: Bangs, TX
Bake Time: 60 minutes
Bake Temperature: 350 degrees

Ingredients:

1/4 Cup margarine
1 8 Oz. pkg. cream cheese
1 Cup sugar
1 egg

1 Cup flour
1 Tsp. baking powder
1/4 Tsp. salt
1 Tsp. vanilla

2 Tbls. sugar
2 Cups blueberries
1 Tsp. cinnamon

Instructions:

Beat margarine and cream cheese at medium speed of an electric mixer until creamy; add 1 cup sugar and egg. Beat well. Combine flour, baking powder and salt; stir into margarine mixture. Stir in vanilla. Fold in berries. Pour into an oblong Pyrex dish, which has been coated with cooking spray. Combine 2 tablespoons sugar and 1 teaspoon cinnamon. Sprinkle over batter. Bake at 350 degrees for 50 minutes to an hour. Cool and cut into bars.

Blueberry Granola Bars

Submitted by: Bill Brooke
From: Beaumont, TX
Bake Time: 40 minutes
Bake Temperature: 350 degrees

Ingredients:

1/2 Cup honey
1 1/2 Cups quick cooking oats

1 Tsp. ground cinnamon
3 Tbls. vegetable oil

1/4 Cup brown sugar
2 Cups fresh blueberries

Instructions:

Preheat the oven to 350 degrees. Coat a 9 by 9 baking pan with cooking spray and set aside. In a medium saucepan combine honey, brown sugar, oil and cinnamon. Bring to a boil over medium heat and continue to boil for 2 minutes. Be sure not to stir. Meanwhile, in a large bowl combine oats and blueberries. Stir in honey mixture until it's well blended. Spread into baking pan and gently press mixture flat. Bake 40 minutes or until lightly brown. Cool completely in the pan or on a wire rack. Cut into bars and enjoy.

Blueberry Sauce

Submitted by: Bill Brooke
From: Beaumont, TX.

Ingredients:

1 orange
1/4 Cup dark seedless raisins

2 1/2 Cups sugar
1 1/2 Cups water

1 lemon
3 Cups fresh blueberries

Instructions:

Wash and dry orange and lemon. With a vegetable peeler remove the outer rind of each, slicing so thinly that none of the white under layer comes with it. Finely chop the rinds. Remove the pulp and chop. Discard seeds. In a large pot bring water and sugar to a boil. Add orange, lemon and raisins. Reduce heat and simmer for 5 minutes. Add the blueberries and cook over medium heat until thick, about 30 minutes. Stir often to prevent sticking. Serve warm over vanilla ice cream.

UTAH

"When my cousin was growing up, their family spent a month of each summer at a family resort in upper New York. We had two favorite activities I remember best. One was sharpening stones into various shapes by the brook and the other was going into the mountains somewhere by Catskills in the morning to pick blueberries. We would find big baskets from the basement to collect the blueberries in. Aunt MaMaee would tell us to try and fill them. Before she made the pies, we would get to have a bowl of fresh blueberries with cream on them. They were excellent."

Blueberry Sour Cream Muffins

Submitted by: Dave Jones
From: Pleasant Grove, UT
Bake Time: 15 minutes
Bake Temperature: 425 degrees

Ingredients:

1/4 Cup butter	1 1/3 Cups flour	3/4 Cup sour cream
3/4 Cup sugar	1/2 Tsp. baking powder	1/2 Tsp. vanilla
2 eggs	1/4 Tsp. salt	1 Cup blueberries

Instructions:

Cream butter and sugar. Beat in eggs, then add sifted dry ingredients and sour cream. Beat in vanilla. Fold in berries. Bake at 425 for 12 to 15 minutes. Makes about a dozen muffins.

Fountain of Youth

Submitted by: Krista Terrelonge
From: Vernal, UT

Ingredients:
1 Cup low-fat cherry yogurt   3/4 Cup frozen blueberries   1/4 Cup cranberry juice
1 Cup frozen unsweetened cherries, pitted

Instructions:
Combine all ingredients in a blender and blend until smooth. Makes 2 1/2 cups. Add ice if desired.

Oma's Blueberry Cobbler

Submitted by: Laura Hall
From: St. George, UT
Bake Time: 50 minutes
Bake Temperature: 350 degrees

Ingredients:
1 Cup flour   1/4 Tsp. salt   1/4 Cup milk
1/4 Cup sugar   1/2 Cup sugar   1/4 Cup butter
1 Tsp. vanilla   1/4 Tsp. ground cinnamon   1 Tsp. baking powder
1 egg   3 Cups blueberries   1 Tsp. butter

Instructions:
Combine flour, 1/2 cup sugar, baking powder and salt in mixing bowl; make well in center. In a large measuring cup, wisk together butter, egg, milk, and vanilla. Pour into well; stir just until evenly moistened. Pour into greased 10 inch pie pan and spread to cover bottom of pan. Toss blueberries, 1/4 cup sugar and cinnamon in bowl. Arrange over dough. Dot with butter. Bake at 350 for 45 to 50 minutes or until juice bubbles up and top is lightly browned.

Blueberry Pizza

Submitted by: Carol Harper
From: Orem, UT
Bake Time: 15 minutes
Bake Temperature: 350 degrees

Ingredients:

1/4 Cup light brown sugar
1 stick of butter
1 Cup flour
1 Cup chopped pecans

3/4 Cup sugar
(1) 8 Oz. pkg. cream cheese
1/2 Pint whipping cream
Raspberries

Fresh blueberries
Peaches
Strawberries
Fresh bananas

Instructions:

Melt butter. Mix flour, brown sugar and pecans, add butter. Pat into pan. Bake at 350 for 15 minutes. Cool. Combine the cream cheese, sugar, and whipping cream. Whip mixture and spread over the crust. Top with the fresh fruit.

Blueberry Fritters

Submitted by: Emma Lou Harris
From: Delta, UT
Bake Time: 3 minutes
Bake Temperature: 375 degrees

Ingredients:

2 Tbls. flour
1 Cup fresh blueberries
1 Cup flour
1/4 Cup sugar

1 egg, beaten
Pinch of salt
2 1/2 Tsp. baking powder

Vegetable oil
1/3 Cup milk
Confectioners' sugar

Instructions:

Combine 2 tablespoons of flour and blueberries, toss lightly. Set aside. Combine 1 cup flour, baking powder, sugar and salt in mixing bowl. Mix well. Combine egg and milk. Gradually add milk mixture to flour mixture, stirring until smooth. Fold in blueberries. Heat 3 to 4 inches of oil to 375 degrees in Dutch Oven. Drop 3 to 4 1/4 cupfuls of batter on at a time. Cook about 3 minutes on each side or until golden brown. Drain on paper towels. Sprinkle the fritters with confectioners' sugar.

Vita Pack

Submitted by: Krista Terrelonge
From: Vernal, UT

Ingredients:

3/4 Cup fresh blueberries   1 Cup low-fat blueberry yogurt
3/4 Cup chilled apple juice   1 Cup frozen peaches, chopped

Instructions:

Combine all ingredients in blender. Blend until smooth. Makes 2 1/2 cups. Add ice if desired.

VERMONT

"After drying berries in the sun, the Indians beat them into a powder and added this powder to parched meal to make a dish called Sautauhig. We found it to be delicious".

Explorer Samuel de Champlain in 1616

Blueberry Chill Pie

Submitted by: Gladys Paddock
From: Bennington, VT

Ingredients:
1/2 Tsp. vanilla
(1) 4 Oz. tub whipped topping
(1) 6 Oz. Keebler Ready Pie Crust Butter Shortbread

1/4 Cup lemon juice
(1) 14 Oz. can sweetened condensed milk

1 can blueberry pie filling

Instructions:
Empty pie filling into strainer to drain excess liquid, reserving some blueberries and liquid for garnish. In a medium bowl, combine condensed milk, lemon juice and vanilla. Mix until well blended. Fold in whipped topping and mix well. Mound 1/3 of mixture into crust and gently spread flat. Spoon a layer of blueberries over mixture. Top with rest of mixture. Freeze until firm, about 3 hours. Let stand at room temperature for 15 to 20 minutes before serving. Garnish with reserved blueberries and liquid if desired.

Blueberry Cake

Submitted by: Myrtle Davidson
From: Northfield, VT
Bake Time: 40 minutes
Bake Temperature: 350 degrees

Ingredients:

1/2 Cup butter
1 1/2 Cups fresh blueberries
1 Tbls. flour
1 Tbls. sugar
2 egg yolks

1 1/2 Cup flour
3/4 Cup sugar
1/4 Tsp. salt
1 Tsp. vanilla
1/4 Cup sugar

1 Tsp. baking powder
1/3 Cup milk
2 egg whites

Instructions:

Cream butter and 3/4 cup sugar. Add salt, vanilla and egg yolks. Beat until creamy. Combine 1 1/2 cup flour and baking powder. Add this alternately with milk to egg yolk mixture. Beat egg whites until soft, adding 1/4 cup sugar, one tablespoon at a time, and beat until whites are stiff. Coat blueberries with 1 tablespoon flour and add to the batter. Fold egg whites in last. Pour into greased 8 inch square pan. Sprinkle top with 1 tablespoon sugar. Sprinkle cinnamon over the sugar. Bake at 350 for 30 to 40 minutes.

Virginia

"Hope you enjoy this recipe. I found it in a magazine on USA Air Flight 898 in May of 1993. I have relatives and friends in CT,TX,KS,AK,AZ,CA,SC, and OR that I will contact for recipes for you…"

"During the Civil War, weary soldiers drank beverages containing blueberries to invigorate themselves after a hard day's work…"

Blueberry Delight

Submitted by: Trisha Knicely
From: Harrisonburg, VA

Ingredients:

2 Cups graham cracker crumbs
1/3 Cup margarine, melted
2 1/2 Cups blueberry pie filling

1/2 Cup sugar
(1) 8 Oz. pkg. cream cheese
1 envelope Dream Whip

1/2 Cup milk
2 Tbls. Sugar

Instructions:

Combine graham cracker crumbs, 2 tablespoons sugar and margarine. Press into bottom of 8 by 12 baking dish. Cream cream cheese and 1/2 cup sugar until smooth. Whip Dream Whip and milk until stiff. Blend with cream cheese mixture. Spread over crumbs and chill until set. Spread filling over top. Creates 6 to 8 servings.

Blueberry Meringue

Submitted by: Edna Green
From: Richmond, VA
Bake Time: 20 minutes
Bake Temperature: 350 degrees

Ingredients:

(2) 12 Oz. pkgs. Frozen blueberries   3 Tbls. grated orange peel   2 Tbls. sugar
4 square or round sponge cakes   3 egg whites

Instructions:

Defrost blueberries. Pour half into a buttered 1 quart baking dish. Lay sponge cake on berries. Cover with remaining blueberries. Sprinkle with orange peel. Whip egg whites stiff with sugar. Cover casserole with meringue. Bake at 300 for 20 minutes or until meringue is golden brown. Creates 4 to 5 servings.

Blueberry Puffs

Submitted by: Brenda LaBauve
From: Charlottesville, VA
Bake Time: 30 minutes
Bake Temperature: 400 degrees

Ingredients:

2 Cups flour, sifted   1/2 Tsp. salt   1/2 Cup milk
6 Tbls. sugar   1 egg   1 Cup fresh blueberries
2 1/2 Tsp. baking powder   1/3 Cup salad oil   2 Tbls. grated lemon

Instructions:

Preheat oven to 400. Sift flour, 4 tablespoons of the sugar, baking powder and salt in a small bowl. Beat egg slightly in a small bowl; stir in salad oil and milk. Add all at once to flour mixture; stir just until mixture is evenly moist; fold in blueberries. Spoon into greased muffin pan cups and fill each 2/3 full. Mix remaining 2 tablespoons sugar and lemon peel in a cup; sprinkle over batter. Make sure to sprinkle the lemon peel and sugar over batter just before putting into oven to keep crisp. Bake 30 minutes. Cool 10 minutes in pan on wire rack. Remove from pan and serve warm. Makes 12 puffs.

Blueberry Custard Pie

Submitted by: Margaret Hill
From: Woodstock, VA
Bake Time: 50 minutes
Bake Temperature: 425 degrees

Ingredients:

2 Cups blueberries
1 9 Inch pie crust, chilled
1 Tbls. Flour

1 Cup sugar
1 Cup evaporated milk
3 eggs, lightly beaten

1 Tsp. vanilla
Cinnamon
Nutmeg

Instructions:

Preheat oven to 425. Place berries in pie shell; set aside. In a medium bowl, mix flour with sugar. Gradually add evaporated milk stirring until smooth. Wisk in eggs and vanilla until blended. Pour mixture over blueberries. Sprinkle with a dash of cinnamon and nutmeg. Bake 15 minutes. Reduce heat to 350 and bake 35 minutes longer.

Blueberry Crunch

Submitted by: Lyn Hoskinson
From: Crewe, VA
Bake Time: 45 minutes
Bake Temperature: 350 degrees

Ingredients:

1 Cup oatmeal
1/2 Cup flour
1/2 Tsp. Salt

1/2 Cup butter
1 Cup brown sugar
1/2 Cup dry milk

1/2 Tsp. Cinnamon
1 1/2 Cups blueberries

Instructions:

Mix oatmeal, brown sugar, flour, milk, salt and cinnamon together. Blend in butter. Spread 2/3 of this mixture in greased 8 by 8 inch baking dish. Over this mixture spread berries. Spread remaining crumb mixture over berries. Bake at 350 for 45 minutes. Serve warm or cold with ice cream or whipped cream.

Blueberry Swirl

Submitted by: Florence Miller
From: Rocky Mount, VA

Ingredients:

1/2 Gallon vanilla ice cream
1 8 Oz. tub whipped topping

1 Quart pureed blueberries

Sugar

Instructions:

Puree blueberries with sugar to taste. Mix other ingredients together and swirl fruit into it. Pour into a 9 by 13 inch dish and freeze.

Back Bumper Blueberry Pound Cake

Submitted by: Regina Meola
From: Dillwyn, VA
Bake Time: 95 minutes
Bake Temperature: 325 degrees

Ingredients:

1 3/4 Cups flour
1 Tsp. baking powder
3/4 Tsp. salt
1/4 Tsp. nutmeg

(1) 6 Oz. pkg. cream cheese
1 stick butter, unsalted
1 3/4 Cups sugar

Grated zest of 1 lemon
4 eggs
1 1/4 Cup blueberries

Instructions:

Preheat oven to 325. Butter and flour a 9 by 5 inch loaf pan, line with wax paper or parchment. Sift flour, baking powder, salt and nutmeg into bowl. Reserve. Using electric mixer, cream butter and cream cheese. Gradually beat in sugar, then eggs, one at a time, beating after each addition. Beat in vanilla and zest. Gently blend dry into creamed mixture in several stages, scraping down sides of bowl as needed. Fold in blueberries. Transfer batter to pan and smooth top. Bake 85 to 95 minutes on center rack, until a tester inserted in center comes out clean. Cool in pan 15 minutes then turn cake out of pan and place on wire rack. Peel off paper and cool. Makes 10 to 12 servings.

Blueberry Gingerbread

Submitted by: Regina Meola
From: Dillwyn, VA
Bake Time: 40 minutes
Bake Temperature: 350 degrees

Ingredients:

1 1/2 Cups blueberries
2 Cups flour
1 Tsp. baking soda
1/4 Tsp. ground mace

1 1/4 Tsp. ground ginger
1 Tsp. ground cinnamon
1 Cup sugar
1/4 Tsp. salt

1/2 Cup vegetable oil
1 Cup buttermilk
1 egg
4 Tbls. molasses

Instructions:

Toss blueberries with 2 tablespoons flour; set aside. Mix remaining flour with soda, ginger, cinnamon, mace, and salt; set aside. In a large mixing bowl beat together oil, sugar, eggs, and dark molasses; mix well. Add flour alternately with buttermilk to creamed mixture; beat until smooth; gently fold in blueberries. Pour into a greased and flour 9 inch square pan. Bake at 350 for 35 to 40 minutes, or until center tests done. Can be served with a dollop of whipped cream or sour cream.

Peek-A-Blue berry farm

WASHINGTON

"My husband and I had a blueberry field for many years. Every year we made a list of recipes to hand out to people who purchased berries. Enclosed is the most popular recipe ever- Blueberry Jell-O Pie…"

Blueberry Mallow Pie

Submitted by: Mary Anderson
From: Seattle, WA

Ingredients:
2 1/2 Cups blueberry pie filling   1/4 Cup milk   1/2 Tsp. vanilla
1/2 Cup heavy cream, whipped   2 Cups mini marshmallows
1 9 Inch graham cracker crust, chilled

Instructions:
Pour pie filling into crust and chill. Melt marshmallows with milk in a double boiler; stir until smooth. Add vanilla, chill until slightly thickened. Mix until well blended. Fold in whipped cream. Spread over pie filling. Chill until firm.

Blueberry Apple Tortillas

Submitted by: Erma Lauser
From: Vancouver, WA

Ingredients:

1 jar blueberry- apple pie filling   1/2 Tsp. ground cinnamon   6 large tortillas
2 Tbls. sugar   2 Tbls. butter

Instructions:

To assemble, brush each tortilla generously with melted butter and sprinkle with cinnamon and sugar mixture. Place 1/6 of the pie filling down center. Fold bottom of the tortilla to partially cover the filling; fold in sides to enclose filling completely. Heat in microwave for 35 seconds. Garnish with additional blueberries and confectioners' sugar.

Blueberry Kuchen

Submitted by: Mrs. Wyatt
From: Bainbridge Island, WA
Bake Time: 60 minutes
Bake Temperature: 400 degrees

Ingredients:

1 Cup flour   2 Tbls. Sugar   1 Tbls. white vinegar
1 Tbls. powdered sugar   2 Tbls. flour   2/3 Cup sugar
3 Cups blueberries   1/8 Tsp. salt   1/2 Cup margarine
1/4 Tsp. cinnamon

Instructions:

If frozen do not defrost blueberries. In a medium bowl, mix 1 cup flour, salt and 1 tablespoon sugar. Cut in margarine until mix resembles a course meal. Sprinkle with vinegar and form into a dough. With lightly floured fingers press the dough into the bottom and up sides of a 9 inch cake pan with removable bottom. The dough should be about 1/4 inch thick on bottom and stand 1 inch high around the sides of the pan. Spread the berries over the dough. Mix the remaining 2 tablespoons flour with the remaining 2/3 cup sugar and cinnamon. Sprinkle this mixture over berries. Bake on the lowest rack of the oven for 50 to 60 minutes or until the crust is well browned and the filling bubbles. Cool; sprinkle with the powdered sugar before serving.

Blueberry Tapioca Pudding

Submitted by: Suellen Brant
From: Longview, WA

Ingredients:

3 Tbls. minute tapioca
1/3 Cup brown sugar
1/4 Tsp. salt

1 Tsp. vanilla
2 1/2 Cups milk
1/2 Tsp. grated orange rind

2 Cups blueberries
4 Tbls. white sugar
2 eggs, separated

Instructions:

Blend tapioca, sugar, and salt together. Beat egg yolks with milk. Gradually stir mixture into tapioca. Cook, stirring constantly until mixture starts to boil. Remove from heat and cool. Stir in vanilla and grated orange rind. Beat egg whites until stiff. Gradually beat in sugar. Fold egg whites into cooled tapioca mixture. Fold in blueberries. Chill. Garnish with whip cream and fresh blueberries. Makes 6 servings.

Zucchini Blueberry Bread

Submitted by: Ruby Taylor
From: Centralia, WA
Bake Time: 60 minutes
Bake Temperature: 350 degrees

Ingredients:

3 eggs
1/4 Tsp. baking powder
1 1/2 Cups sugar or honey
3 Tsp. vanilla

1 Tsp. salt
1 Tsp. baking soda
1 Cup oil
3 Tsp. cinnamon

2 Cups ground zucchini
1 Cup chopped nuts
3 Cups flour
2 Cups blueberries

Instructions:

Beat eggs until fluffy. Add sugar, vanilla, and oil. Blend well. Stir in zucchini and fold in blueberries. Sift in flour, baking powder, soda, cinnamon and fold in nuts. Bake in 2 greased log pans at 350 for an hour.

Blueberry Roll

Submitted by: Corrine Gunderson
From: Cle Elum, WA
Bake Time: 45 minutes
Bake Temperature: 400 degrees

Ingredients:

1 1/2 Cups flour
2 Tsp. baking powder
1/2 Tsp. salt

2 Tbls. Sugar
1 egg
5 Tbls. shortening

1/4 Cup milk
1 1/2 Cups blueberries
1/2 Cup sugar

Instructions:

Sift flour, baking powder, salt, and 2 tablespoons sugar in a bowl. With pastry blender, cut in shortening until mixture resembles a coarse corn meal. With fork, beat eggs slightly in measuring cup and add milk to make 1/2 cup in all; blend well. Add to flour mixture and stir quickly with a fork until just mixed. Turn onto lightly floured board; with fingertips, gently knead 10 times until outside looks smooth. Roll or pat into rectangle about 8 by 10 inches. Cover dough with blueberries except for 1 inch on both sides. Sprinkle the 1/2 cup sugar over berries. Starting at one long side, roll up like a jellyroll. Place in greased shallow baking pan, about 11 by 7 inches. Bake at 400 for 40 to 45 minutes, or until golden brown. Cut into slices and serve hot. Add a fluff of cream, whipped until just thick; sweeten slightly and sprinkle with nutmeg. Can also use a hard or lemon sauce..

Blueberry Eggnog Pie

Submitted by: Corrine Gunderson
From: Cle Elum, WA

Ingredients:

1/2 Cup heavy cream, whipped
(1) 9inch baked pie shell
1/4 Tsp. salt
1 1/2 Cups milk

1/3 Cup sugar
1 Tbls. unflavored gelatin
1/4 Cup sugar
Sugared blueberries or blueberry sauce

3 egg whites
1 Tsp. vanilla
3 beaten egg yolks

Instructions:

In a double boiler, mix the 1/3 cup sugar, gelatin, and salt. Add egg yolks and milk. Cook and stir over hot, not boiling, water until mixture is slightly thick. Remove from heat, add vanilla. Chill, stirring occasionally, until mixture mounds slightly when spooned. Beat egg whites to soft peaks. Gradually add 1/4 cup sugar, beating to stiff peaks. Beat gelatin mixture just until smooth; fold into egg whites. Fold in whipped cream. Pile into cooled pastry shell (have edges crimped high). Chill firm. Serve with blueberry sauce on top of blueberries.

Blueberry Peek-A-Boos

Submitted by: Barbara Ingrum
From: Bothell, WA
Bake Time: 15 minutes
Bake Temperature: 450 degrees

Ingredients:

1 Cup fresh blueberries
4 Tbls. sugar
2/3 Cup milk
1 Cup confectioners' sugar

2 Cups Bisquick
1/4 Cup butter
1/2 Tsp. cinnamon
1 1/2 Tbls. cream or until easy to spread

1 Tsp. vanilla
1/2 Tsp. nutmeg
2 Tsp. lemon juice

Instructions:

Heat oven to 450. Toss blueberries, 2 tablespoons sugar, nutmeg, cinnamon and lemon juice. Set aside. Combine Bisquick, 2 tablespoons sugar and butter. Add milk, all at once; stir with a fork into soft dough. Beat 20 strokes. Place a tablespoonful of dough in the bottom of 10 paper lined muffin cups. Top with 1 tablespoon of blueberry mixture. Drop a tablespoon of dough on top of the berries. Bake 10 to 15 minutes, or until golden brown. Remove from oven. When slightly cool, frost with a mixture of powdered sugar, vanilla and cream.

Blueberry Clafouti

Submitted by: Josephine Hadfield
From: Bremerton, WA
Bake Time: 60 minutes
Bake Temperature: 350 degrees

Ingredients:

3 Cups blueberries
Confectioners' sugar
1 1/2 Tbs. flour
1 Tsp. vanilla

1/2 Cup sugar
3/4 Cup heavy cream
4 Tbls. sugar
3 eggs

1 Cup flour
Pinch of salt
1/2 Cup milk

Instructions:

Use 1 1/2 tablespoon butter and 4 tablespoons sugar to butter and sugar a 2 quart fireproof baking dish. Place berries in dish. Place cream, milk, sugar, vanilla, salt and eggs in mixing bowl and process for a few minutes to mix well. Add flour and process until just smooth, about 3 seconds. Pour batter over berries, place in oven and bake for about an hour or until puffed and browned. The clafouti is done when a knife can be inserted in the center and come out clean. Sprinkle with powdered sugar and serve warm. This will sink as it cools. Makes 6 to 8 servings.

Blueberry Rum Crème

Submitted by: Jim Claire
From: Bremerton, WA

Ingredients:

2 Cups blueberries   1/4 Cup sugar   2 Cups milk
1 Cup whipping cream   1/2 Tsp. rum extract
1 small pkg. instant vanilla pudding

Instructions:

Combine blueberries and sugar. Whip cream until stiff. Prepare pudding according to directions. Stir in rum extract. Allow pudding to set 5 minutes. Fold in whipped cream and blueberries. Chill in individual serving dishes. Serves 6 to 8.

Blueberry Jell-O Pie

Submitted by: Margaret Freeman
From: Puyallup, WA
Bake Time: 30 minutes
Bake Temperature: 375 degrees

Ingredients:

(1) 3 Oz. pkg. raspberry Jell-O   1/2 Cup sugar   Cornstarch
4 Cups blueberries   (1) 9 Inch pie shell, double crust   1/2 Cup water

Instructions:

Bring berries, water and sugar to a light boil, add Jell-O and enough cornstarch to thicken. Pour into pie shell. Bake at 375 for 25 to 30 minutes.

WEST VIRGINIA

"We are from West Virginia, the South Eastern corner, Monroe County. It is beautiful here right in the mountains. This past winter, our family visited in New York in your area, visiting our Amish relatives. Stop down and visit if you are ever down this way- we have a bakery and make lots of goodies..."

Blueberry Milkshake

Submitted by: Doris Schwartz
From: Gap Mills, WV

Ingredients:
3 Cups frozen blueberries   1/4 Cup brown sugar   Milk

Instructions:
In a 5 cup blender, put blueberries. Add brown sugar. Add milk to cover berries. Blend. Add more milk if it's too thick. Makes about 2 tall glasses. Serve right away.

Blueberry Muffins

Submitted by:　　　Jennifer Stewart
From:　　　　　　Buffalo, WV
Bake Time:　　　　25 minutes
Bake Temperature:　375 degrees

Ingredients:

4 Cups flour
2 1/2 Cups milk
2 Cups sugar
1/2 Tsp. Cinnamon
3 Cups blueberries
4 Eggs

1/4 Tsp. ground cloves
4 Tsp. baking powder
2 Tsp. vanilla
1 Tsp. salt
2 Tbls. sugar

1 Cup butter
1/2 Tsp. baking soda
1/8 Tsp. ginger
Juice of 1/2 a lemon
1/2 Tsp. nutmeg

Instructions:

Combine 4 cups flour, baking powder, ginger, cloves, cinnamon, baking soda and salt. Cream together butter and 2 cups sugar. Add eggs, milk, lemon juice and vanilla. Stir in dry ingredients by hand just until moistened. Fold in blueberries. Fill greased of paper-lined muffin cups 2/3 full. Combine 2 tablespoons sugar and nutmeg. Sprinkle over muffins. Bake at 375 for 20 to 25 minutes or until muffins test done. Make sure to understir to prevent toughness. For a different twist replace 1/2 cup milk with 1/2 cup plain yogurt.

Blueberry No-Bake Ice Cream Pie

Submitted by:　　　Marion Huffman
From:　　　　　　Wheeling, WV

Ingredients:

3 Cups blueberries
3/4 Cup sugar
1 1/2 Tbls. lemon juice
(1) 9 inch ready made graham cracker crust

2 1/2 Tbls. cornstarch
2 Tbls. butter
1 Quart vanilla ice cream

2 Tbls. orange juice
1/4 Cup water

Instructions:

Combine blueberries and sugar in a 1 1/2 quart microwave able dish. Mix water and cornstarch; stir into blueberries. Cover and vent. Microwave on high for 5 to 8 minutes, stirring after 3 minutes, or until mixture is thickened and bubbly. Stir in butter and lemon juice. Cool to room temperature. Reserve 1/2 cup filling and refrigerate. Pour filling into graham cracker crust. Cover with plastic wrap. Place in freezer at least 2 hours to firm. To soften ice cream, place in a 2 quart microwave able bowl. Microwave on medium for 1 to 1/4 minute, stirring after 30 seconds. Spoon ice cream over firmed blueberry filling. Freeze. Remove pie 30 minutes before serving to thaw enough to cut. Combine reserved filling and orange juice. Cut into pie wedges and pour sauce over each wedge before serving. Makes one 8 inch pie.

Wisconsin

" Here is a recipe for blueapple pie that was world premiered at The Blue Hill Fair-1997, Yummy…"

" As this is a letter from Wisconsin, I will tell you about our city, Pittsville. It is very small-less than 900 population. As you may guess, years ago it was much larger. We are known for being " The Exact Center of Wisconsin". The river running through town has a small island and a stake was put there many years ago…"

" I live on a farm in Wisconsin that has been in my family for 80 years. I have a Beta Fish as a pet. I call the fish Norton…"

Blueapple Pie

Submitted by: Mary Maeder
From: West Bend, WI
Bake Time: 60 minutes
Bake Temperature: 350 degrees

Ingredients:

1 bag frozen blueberries (8 to 12 oz.)
1 Cup sliced apple
1 Tbls. instant tapioca mix

1/4 Tsp. nutmeg
1/4 Tsp. cinnamon

2 frozen piecrust shells
1/3 Cup sugar

Instructions:

Place apples into piecrust. Mix tapioca, sugar and spices with blueberries and pour over the apples. Fill to the brim but do not overfill. Seal the top crust as tight as possible and poke a few fork holes. Bake at 350 for 50 to 60 minutes. May vary ingredients to suit taste and your pie plate size. Double this recipie for 9" pie.

Blueberry Marmalade

Submitted by: LouAnn Przesmicki
From: Wausau, WI

Ingredients:

1 medium sized orange
1 medium sized lemon
1/8 Tsp. soda

3 Cups blueberries, crushed
5 Cups sugar

1 bottle fruit pectin
3/4 Cup water

Instructions:

Remove peel in quarters from lemon and orange. Lie quarters flat and remove about half the white part. With a sharp knife, cut remaining peel into shreds. Place in a large kettle. Add water and soda; bring to a boil and simmer covered 10 minutes; stirring occasionally. Chop peeled fruit and add pulp and juice to undrained peel. Cover and simmer 15 minutes. Stir in crushed blueberries and sugar. Bring to a boil. Cook gently 5 minutes. Remove from heat and stir in fruit pectin. Stir and skim by turn for 5 minutes. Pour into sterilized glasses. Paraffin at once. Makes about eight 6 ounce glasses.

Fruit Salad with Blueberry Cheese Dressing

Submitted by: LouAnn Przesmicki
From: Wausau, WI

Ingredients:

6 slices canned pineapple
crisp lettuce leaves
2 Cups melon balls

(1) 3 Oz. pkg. cream cheese
1 Cup blueberries
2 Tbls. mayonnaise

2 Tsp. Sugar
1 Tsp. Lemon juice
1/8 Tsp. Salt

Instructions:

Arrange pineapple slices on lettuce leaves. Top with melon balls and 1/2 cup blueberries. Blend mayonnaise and cream cheese together. Add remaining blueberries, sugar, lemon juice and salt. Beat with a rotary beater until well blended. Makes 2/3 cup dressing. Salad serves 6.

Blueberry Peach Pound Cake

Submitted by: Laura Ritter
From: Reeseville, WI
Bake Time: 70 minutes
Bake Temperature: 350 degrees

Ingredients:

1/2 Cup butter
1 1/2 Cup sugar
3 eggs
1/4 Tsp. salt

1/2 Cup milk
2 1/4 Cups chopped peaches
2 Tsp. baking powder

Confectioners' sugar
2 1/2 Cups cake flour
2 Cups blueberries

Instructions:

Cream butter and sugar; beat in eggs, one at a time. Beat in milk. Combine flour, baking powder and salt. Add to creamed mixture. Fold in peaches and blueberries. Pour into greased and floured 10 inch fluted tube pan. Bake at 350 for 60 to 70 minutes, or until tester comes out clean. Cool in pan 15 minutes, remove to a wire rack to cool completely. Dust with powdered sugar. Makes 10 to 12 servings.

Blueberry Apple Crisp

Submitted by: Lynn Shallberg
From: Brookfield, WI
Bake Time: 45 minutes
Bake Temperature: 375 degrees

Ingredients:

6 Cups sliced apples
1/2 to 2/3 Cup sugar
1/2 Cup light brown sugar

1 Cup flour
1/2 Cup melted butter

1 to 2 Cups blueberries
Cinnamon

Instructions:

Mix apples and blueberries with sugar and place in 8 by 8 baking dish. Sprinkle cinnamon over fruit. Mix flour, brown sugar and melted butter. Pour over blueberry mixture. Bake at 375 for 45 minutes. Can be served with ice cream or whipped cream.

Brandy Blue Cake

Submitted by: Audrey Halverson
From: Ellsworth, WI
Bake Time: 75 minutes
Bake Temperature: 325 degrees

Ingredients:

1 Cup floured blueberries   5 eggs   1/2 Tsp. mace
1/2 Lb. Butter   2 Oz. brandy   1 2/3 Cups sugar
2 1/4 Cups cake flour - sift before measuring.

Instructions:

Cream butter and sugar. Add eggs and beat, then add flour, brandy and mace. Fold in floured blueberries. Cook in a well greased and floured angel cake pan. Bake at 325 for 1 hour and 15 minutes.

Blueberry Drink Base

Submitted by: Audrey Halverson
From: Ellsworth, WI

Ingredients:

1 Quart fresh blueberries   1 Cup sugar   2 Cups water

Instructions:

Cook together berries, sugar and water. Strain and store juice in refrigerator. Add two tablespoons to a glass of lemonade. To create Blueberries on the Rocks, dash in Angostura bitters to drink base over ice.

WYOMING

"I am an inmate at the Wyoming State Penitentiary and also work in the bakery here making breads and desserts for about 600 guys. We use lots of blueberries. We often make blueberry muffins and add blueberries to pancakes just for us 15-20 kitchen workers. I personally like to add blueberries to hot cereal like oatmeal…"

"Here in Wyoming there are very few roadside stands but many dude ranches where paying guest enjoy Western hospitality. At some dude ranches, the guests even help with the ranch chores…"

"I was inspired to write to you today, not only because this is a great blueberry recipe, but also because my dog is named Blueberry. I named her this because I thought she looked like blueberry muffin or pancake batter… I'm sure you will have no problem getting a recipe from every state, but now you can say that you have one from a cattle ranch in Wyoming."

Quick Fruit Salad

Submitted by: Edith Shepherd
From: Moorcroft, WY

Ingredients:

Grapes   Chopped nuts   Apples
Blueberries   Mandarin oranges   1 can blueberry pie filling
Bananas   Pineapple

Instructions:

Prepare individual fruits in bowl. Toss together lightly. Mix in pie filling. Chill. For extra large amount, use two cans of filling.

Blueberry 'N' Spice Sauce

Submitted by: Irma Bideau
From: Casper, WY

Ingredients:
1/2 Cup sugar
1 Tbls. cornstarch
1/4 Tsp. cinnamon
1/4 Tsp. nutmeg
1/2 Cup hot water
2 Cups blueberries

Instructions:
In a small saucepan combine sugar, cornstarch, cinnamon and nutmeg; gradually stir in water. Cook, stirring constantly, over low heat until mixture thickens and comes to a boil. Add blueberries and continue stirring until sauce reaches a boil again. Reduce heat, simmer 5 minutes. Serve warm over ice cream or cake.

Blueberry Lemon Cheesecake Pie

Submitted by: Ruth Uptain
From: Casper, WY
Bake Time: 10 minutes
Bake Temperature: 450 degrees

Ingredients:
1 can blueberry pie filling
1 to 2 Tsp. grated lemon peel
1/2 Cup whipping cream, whipped
(1) 8 Oz. pkg. cream cheese
(1) 15 Oz. pkg. refrigerated piecrusts
1/3 Cup sugar

Instructions:
Heat oven to 450. Prepare piecrust according to directions for one crust baked shell using 9 inch pan. Bake for 9 to 11 minutes or until lightly browned. Cool. In a small bowl, beat cream cheese, sugar and lemon peel until light and fluffy. Fold in whipped cream. Spoon into cooled piecrust; spread evenly. Spoon blueberry filling evenly over lemon filling. Refrigerate at least 1 hour before serving. Garnish with whipped cream. Store in refrigerator. Makes 8 servings.

Blueberry Cobbler

Submitted by: Necel Golden
From: Weston, WY
Bake Time: 25 minutes
Bake Temperature: 375 degrees

Ingredients:

1 stick butter
1 Tbls. baking powder
(1) 20 Oz. can blueberry pie filling

1 Cup milk
1 Cup sugar

1 Cup flour
1 pinch salt

Instructions:

Preheat oven to 375. Melt butter in large baking pan (9 by 13). In mixing bowl, combine flour, sugar, baking powder, salt and milk. Pour batter into pan of melted butter. Spoon pie filling over batter. Bake at 375 degrees for 25 minutes. Sprinkle top with sugar, if desired. Best served warm with vanilla ice cream.

Peek-A-Blue berry farm

INTERNATIONAL

"This dessert was created for the famous ballerina Anna Pavlov when she came to dance in Australia in the 1950's. The dessert is called a Pavlova. It can be decorated with any summer fruit but I like to decorate it with blueberries, strawberries and kiwi fruit…"

Submitted from Australia

"I thought you might like to have a recipe from far far away. I used to make the one enclosed from wild blueberries picked near where I live in what is known as the "Cariboo Country" of British Columbia, Canada's westernmost province…."

Blueberry Nut Bread

Submitted by: La Vern Kallinen
From: Australia
Bake Time: 60 minutes
Bake Temperature: 350 degrees

Ingredients:

2 eggs
1 Cup sugar
2 Tbls. melted shortening
4 Tsp. baking powder

1/2 Tsp. salt
1 Tsp. vanilla
3 Cups flour

1/2 Cup chopped nuts
1 Cup blueberries
1 Cup milk

Instructions:

Beat eggs. Gradually add sugar then beat in shortening, milk and vanilla. Add flour, salt, and baking powder. Stir only until mixed. Fold in blueberries and nuts. Pour into a well-greased loaf pan. Bake at 350 degrees for 1 hour. Serve warm with ice cream or butter or cool completely and enjoy at room temperature.

Blueberry Pandowdy

Submitted by:	Audrey Marion
From:	Canada
Bake Time:	25 minutes
Bake Temperature:	375 degrees

Ingredients:

4 Cups blueberries
1/2 Cup milk
1/2 Cup sugar
1 egg

1 1/2 Cup flour
2/3 Cup sugar
2 Tbls. lemon juice

1/2 Cup butter
2 Tsp. baking powder
1/2 Tsp. salt

Instructions:

Combine blueberries, 2/3 cup sugar, and lemon juice. Spread in a 9 by 9 inch pan. Mix flour, baking powder, and salt. Set aside. Cream butter and 1/2 cup sugar. Add egg and beat. Add flour mixture alternately with milk and beat. Pour on top of blueberry mixture. Bake at 375 degrees for 25 minutes or until lightly browned. Cut into squares. Serve warm with cream.

Pavlova with Blueberries

Submitted by:	Caroline Keeler
From:	Australia
Bake Time:	275 degrees
Bake Temperature:	2 hours

Ingredients:

4 egg whites at room temperature
8 Oz. caster sugar
1 Tsp. white wine vinegar

1 Pint strawberries
1 Tsp. corn flour (not cornmeal)
1 1/3 cup double cream

Blueberries
1-2 Kiwi fruit

Instructions:

Preheat oven to 275 degrees. Line a wooden or metal baking tray with parchment paper. Mark on it an 8 inch circle and set aside. Place egg whites in a mixing bowl and wisk until stiff, add the caster sugar a little at a time continuing to whisk until stiff and shiny. Add the wine vinegar and corn flour; stir briefly. Spread the mixture on the drawn circle to cover it gently. Smooth sides and top. Transfer to a preheated oven and bake for 2 hours. Allow cooling completely by turning off the oven and opening the oven door. This will help reduce cracking. When pavlova has completely cooled, remove from parchment paper and transfer to a serving plate. The pavlova will be harder on the outside with a soft marshmallow like center.

Blueberry Dessert Pizza

Submitted by: Emily Boomer
From: Germany
Bake Time: 20 minutes
Bake Temperature: 350 degrees

Ingredients:

1 pkg. white cake mix
1/2 Tsp. cinnamon
1/2 Cup chopped nuts

2 Cups blueberry pie filling
1/2 Cup butter
1 1/4 Cup quick cooking rolled oats

1/4 Cup brown sugar
1 egg

Instructions:

Heat oven to 350 degrees. Grease 12 inch pizza pan or 13 by 9 pan. In a large bowl, combine cake mix, 1 cup oats and 6 tablespoons butter at low speed until crumbly. Reserve 1 cup crumbs for topping. To remaining crumbs blend in egg. Press in prepared pan. Bake at 350 for 12 minutes. In the same large bowl, add remaining 1/4 cup oats, 2 tablespoons butter, nuts, sugar and cinnamon, to reserve crumbs; mix well. Remove base from oven and spread pie filling evenly over top. Sprinkle with reserved crumb mixture. Return to oven and bake 15 to 20 minutes or until crumbs are light golden brown. Cool completely. Cut into wedges or squares.

Blueberry Mousse in White Chocolate Cups

Submitted by: Audrey Marion
From: Canada

Ingredients:

6 squares white chocolate (6 Oz.)
1 Tbls. shortening
2 Egg whites beaten with 2 Tbls. sugar

1 1/2 envelopes unflavored gelatin
1/2 Cup sugar

2 Cups blueberries
1 Cup whipping cream

Instructions:

Melt white chocolate and shortening in top of double boiler over hot water. Remove from heat but keep over warm water. Spread thin layer of chocolate inside of 12 pleated foil cupcake lines that have been placed in muffin pan cups. Refrigerate for 1 hour, and coat again. Refrigerate. When thoroughly hardened, peel foil from chocolate carefully and refrigerate until ready to fill. Puree blueberries in blender or food processor. Transfer to saucepan. Sprinkle gelatin and 1/2 cup sugar over top. Place over low heat and stir to dissolve gelatin and sugar. Place over iced water, stirring often until mixture begins to thicken. Beat egg whites until foamy. Add in 2 tablespoons sugar and beat to form soft peaks. In another bowl, beat whipping cream until stiff. Fold whipping cream into blueberry mixture then fold in beaten egg whites. Fill chocolate cups with blueberry mixture. Garnish with some fresh blueberries and refrigerate until serving.

Blueberry Flan

Submitted by: Audrey Marion
From: Canada
Bake Time: 15 minutes
Bake Temperature: 400 degrees

Ingredients:

2 pkgs. cream cheese
1/2 Cup butter
1 1/2 Cups flour

1 Cup cream (whipped)
1/8 Tsp. salt
1/3 Cup sugar

1 Tsp. lemon juice
Blueberries

Instructions:

Combine 1/2 package cream cheese, butter, flour, and salt. Press into flan pan. Bake at 400 degrees for 12 to 15 minutes. Let cool. Beat together the rest of the cream cheese, cream, sugar, and lemon juice. Spread on top of flan. Top with fresh glazed blueberries.

Blueberry Pork Chops

Submitted by: Audrey Marion
From: Canada

Ingredients:

2 Tbls. brown sugar
2 Tbls. Dijon mustard
4 center-cut pork chops or 8 boneless center-cut pork slices

1 Cup blueberries
Salt and pepper

Vegetable oil

Instructions:

Heat a frying pan and sauté the pork chops in a small amount of oil until browned and cooked through. Salt and pepper the chops and transfer them to a platter and keep them warm in the oven. While the chops are cooking, mix together the blueberries, mustard and sugar. After the chops are cooked, remove excess oil from the frying pan. Add the blueberry mixture to the pan and bring to a boil, scraping up any brown bits. Cook until the sauce is slightly thickened. Drain any juices that have accumulated around the chops. Pour the sauce around the chops and serve.

Blueberry Maple Squares

Submitted by: B. Peterson
From: Canada
Bake Time: 35 minutes
Bake Temperature: 350 degrees

Ingredients:

2 Cups sifted flour
1/3 Cup sugar
2 eggs

1 Tbls. baking powder
1 Cup milk
1/2 Tsp. salt

2 Cups blueberries
2 Tsp. maple flavoring
1/3 Cup melted butter

Instructions:

Mix dry ingredients. Beat eggs, milk, butter and maple flavoring. Add to dry mixture. Stir gently. Fold in blueberries. Spread in a well-greased square pan. Bake at 350 degrees for 35 minutes. Cut into 3 inch squares. Serve warm with dabs of butter and honey.

Blueberry Relish

Submitted by: Gertrude Pare
From: Canada

Ingredients:

4 Cups blueberries

1 medium sized lemon

2 Cups sugar

Instructions:

Coarsely chop the blueberries. Cut lemon into quarters. Remove all seeds but retain the peel. In a blender, chop the lemon finely. Add the sugar and mix together. Creates 4 cups. Can be served on it's own or with grilled chicken or ham, even ice cream.

Blueberry Cream Squares

Submitted by: Dorothy Wheeler
From: Canada
Bake Time: 10 minutes
Bake Temperature: 350 degrees

Ingredients:

2 Cups crushed vanilla wafers
1/4 Cup melted butter
2 Tbls. confectioners' sugar
1 1/2 Cups mini marshmallows

3 Cups blueberries
3/4 Cup sugar
1/4 Cup cornstarch

1 Cup cream, whipped
1 Tsp. lemon juice
1/4 Cup water

Instructions:

Combine 1 cup crushed wafers and melted butter. Place in a buttered 8 by 8 inch pan. Bake at 350 degrees for 10 minutes, remove and cool. Combine cornstarch, sugar and water in saucepan. Add berries. Cook over low to medium heat, stirring until thickened, about 10 minutes. Add lemon juice and cool. Whip cream. Blend in powdered sugar and marshmallows. Spread on crumb crust. Cover with cooled blueberry filling and top with remaining vanilla crumbs. Makes 9 servings.

INDEX

INDEX - Recipes by Category

Pies

Muffins-Breads

Cobblers-Crisps

Cakes

Sauces-Syrups-Jams-Jellies

Pancakes-Breakfast

Drinks

Desserts

108	Iced Blueberry Dessert	Indiana
110	Summer Dessert Pizza	Indiana
111	Blueberry Pudding	Iowa
113	Blueberry Tortilla Pizza	Iowa
117	Blueberry Streusel	Kansas
119	Blueberry Turnovers	Kentucky
121	Blueberry Torte	Kentucky
121	Blueberry Surprise	Kentucky
122	Blueberry Strudel	Kentucky
123	Upside Down Blueberry Pudding	Louisiana
127	Wild Blueberry Trifle	Maine
130	Blueberry Ice Cream	Maine
134	Blueberry Dumplings	Maryland
137	Blueberry Gingerbread	Massachusetts
138	Blueberry Apple Betty	Massachusetts
139	Blueberry Betty for 50	Massachusetts
142	Blueberries in the Snow	Michigan
149	Blueberry Boy Bait	Minnesota
151	Creamy Frozen Fruit Cups	Minnesota
158	Blueberry Ambrosia	Mississippi
158	Honeyed Blueberry Sherbet	Mississippi
161	Blueberry Ice Cream	Missouri
163	Blueberry Raspberry Tart	Missouri
165	Blueberry Bread Pudding with Spiced Sauce	Missouri
167	Blueberry Green Grape Dessert	Montana
171	Frozen Blueberry Dessert	Nebraska
172	Blueberries in the Snow	Nebraska
174	Blueberry Treat	Nebraska
177	Blueberry Butterscotch Russe	New Hampshire
178	Blueberry Fungi	New Hampshire
181	Blueberry Sorbet	New Jersey
182	Blueberry Gingerbread	New Jersey
186	Lemon Blueberry Salad	New Mexico
188	Blue Ambrosia	New York
189	Blueberry Pizza	New York
189	Pumpkin Cake Roll	New York
190	Blueberry Flummery	New York
193	Individual Blueberry Cheesecake	North Carolina
196	Blueberry Ice Cream	North Carolina
201	Blueberry Salad	North Dakota
207	Blueberry Applesauce Bread	Ohio
208	Blueberry Topped Sponge Flan	Ohio
210	Diabetic Blueberry Salad	Oklahoma
214	Blueberry Doughnuts	Oregon
216	Blueberry Charlotte Russe	Oregon
217	Blueberry Ice Cream Puffs	Oregon

C<small>ONTACT</small>

E-Mail: PeekABlueberry@aol.com

Address:
Peek-A-Blue Berry Farm
5555 Oregon Rd.
Bath, New York 14810

Web site: PeekABlueberry.Bookscapes.com